NEURAL NETS AND
CHAOTIC CARRIERS

2ND EDITION

T0324817

Advances in Computer Science and Engineering: Texts

Editor-in-Chief: Erol Gelenbe *(Imperial College)*
Advisory Editors: Manfred Broy *(Technische Universitaet Muenchen)*
Gérard Huet *(INRIA)*

Advances in Computer Science and Engineering: Texts   Vol. 5

NEURAL NETS AND CHAOTIC CARRIERS

2ND EDITION

Peter Whittle

University of Cambridge, UK

Imperial College Press

ICP

Published by

Imperial College Press
57 Shelton Street
Covent Garden
London WC2H 9HE

Distributed by

World Scientific Publishing Co. Pte. Ltd.
5 Toh Tuck Link, Singapore 596224
USA office: 27 Warren Street, Suite 401-402, Hackensack, NJ 07601
UK office: 57 Shelton Street, Covent Garden, London WC2H 9HE

British Library Cataloguing-in-Publication Data
A catalogue record for this book is available from the British Library.

NEURAL NETS AND CHAOTIC CARRIERS (2nd Edition)
Advances in Computer Science and Engineering: Texts — Vol. 5

ISBN-13 978-1-84816-590-8

Desk Editor: Tjan Kwang Wei

Typeset by Stallion Press
Email: enquiries@stallionpress.com

Printed in Singapore by World Scientific Printers.

Preface to the Second Edition

This work seeks to develop and integrate two themes. One is that of associative memory, a topic well worked over, but still needing a development from clear principles. The other is the particular theory of oscillatory networks presented in a highly significant series of papers by W.J. Freeman and his collaborators over more than twenty years; see the references in Chapter 11.

The concern is to investigate artificial neural networks (ANNs) rather than biological neural networks (BNNs) — a stance which absolves a non-biologist of presumption while leaving him full freedom. However, as is usual, one looks to physiology for both inspiration and confirmation; hoping also to repay the debt by the development of some insight from the idealised ANN models.

Part II of the text attempts to develop rational designs for an associative memory from some set of principles. Such principles must stem from an understanding of the function which is required, and this comes first in Chapter 8, on 'fading data'. The function of a full associative memory is to form an inference on the basis of data which may be fleeting and which cannot be stored without degradation. The dynamics of the memory must then be such as to hold significant aspects of the data while the inference is being formed. Quantisation, in some sense, is necessary.

A probability-maximising principle, which turns out to have optimality properties, is adopted to achieve this. It yields an appealing solution: a dynamic, autoassociative and somewhat fuzzy form of the Hamming net. However, the equivalent structure turns out to have real interest, particularly when the images represented by the data are compound in some sense. The possibility of compounding implies a structure which we regard (see Chapter 14) as a fair and insightful model of the transmission/processing loop constituted by the olfactory bulb and the anterior olfactory nucleus in the olfactory system. Of course, one must beware of facile identifications, but the two structures show striking and unforced correspondence on a number of significant features. We find the correspondence striking enough that we venture a couple of testable predictions; see section 14.5.

This investigation also clarifies a number of side issues; e.g. it explains how the 'spurious' equilibria of the Hopfield net have virtually been designed into it.

Readers who wish to follow just this associative-memory theme can probably get away with reading Chapters 5, 6, 8 and 14. Readers determined to plunge *in medias res* should make what they can of Chapter 14.

The other theme is that of oscillatory operation, so evidently a feature of biological neural systems. This must convey some advantage, the obvious one being that, since absolute activity levels are almost meaningless in the biological context, information must be transmitted by a variable rather than a static signal. An oscillatory signal can then be seen as a way of transmitting information despite a background of both noise and irrelevant slow variation in baseline activity. However, one-way 'transmission' is far too naive a concept; biological systems are locked into a dynamic whole by both forward and backward connections, and achieve communication by entering a joint dynamic state.

No approach to these ideas has been more insightful and thorough than that of Freeman and his school, which has combined anatomical observation and physiological experiment with hard thinking and mathematical analysis. In this way Freeman has elucidated a number of fundamental mechanisms and deduced very convincing models (notably for the olfactory system) tested by analogue simulation.

However, while the Freeman theory produces a physiologically faithful model for an oscillator which has certain quite specific properties as a threshold element, and shows how these can be coupled together to form systems of the character and behaviour which are observed, the particular structures suggested for the associative-memory function do not go beyond a Hopfield net. Again, design principles are needed.

In Part III we try to supply these by incorporating the Freeman mechanisms into the nets derived in Part II. Both sets of models are dynamic, but the static equilibria of the Part II models are now replaced by dynamic equilibria. The combination of oscillatory units and the global standardising operation inherent in the Hamming net has a remarkable consequence: the generation of a slow square-wave global oscillation, the 'escapement oscillation', which explains the occurrence of gamma-wave bursts and has a clearly synchronising effect.

Readers who wish to follow just the oscillation theme can read Chapters 11 and 12 and much of Chapter 14 directly, but can probably not read Chapter 13 and complete Chapter 14 without cutting back to Chapters 5 and 8.

Finally, we should clarify the reference to chaos in the title. Any electroencephalogram has a truly chaotic look about it, but we regard systems as operating in spite of chaos rather than in virtue of it. It is a matter of observation that the major components of the olfactory system show relatively simple behaviour in isolation — either a stable equilibrium or a limit cycle — but that when coupled together they show the complex dynamics which one would term chaotic. This is true both of the actual organism and of Freeman's electronic analogue. The appearance of chaos seems to be a consequence of complexity plus transmission delays in the total system. However, the point, made well and early by Freeman, is that it does not matter. If the system is locked into a synchronised whole then communication is achieved by the sensing of the joint dynamic metastate.

I am most grateful to Colin Sparrow, who was not only good enough to carry through the calculations of Chapters 11–13, but also provided steady expertise and perceptiveness. The launch of the first edition in 1998 was marred by a marketing miscalculation; I am most grateful to Imperial College Press for giving the venture a second and better chance. I have made no major changes to the text but, on reading it after this interval of time, am convinced that the reader would do well to follow the short-cuts suggested above. That is, that a reader interested primarily in the associative memory theme could well proceed directly to Chapters 5, 8 and 14; one interested primarily in the oscillation theme could proceed directly to Chapters 11, 12 and 14, but then supplement these by a return to Chapters 5, 8 and 13.

The three graphs of the cover design represent, in descending order: a voltage trace from the olfactory bulb of a rat, a trace from Freeman's electronic simulator of the olfactory system, and a trace from a two-stage Freeman oscillator with feedback and dynamic stabilisation of signal strength.

Contents

Part I

Opening and Themes

Chapter 1

Introduction and Aspirations

Readers who like to skip introductions are advised to read the preface, which also sets out our two central aspirations. One is to develop rational designs for an associative memory from a clear set of principles; principles which must in turn stem from an understanding of required function. The other, seemingly unrelated, is to study the role and effect of oscillatory operation in neural networks, leaning heavily upon the many insights brought to this topic by W.J. Freeman. An understanding of rational design is at least as necessary for this case as for any other; hence the connection.

Neural nets, whether biological or artificial, are concerned with the calculations that go on inside a system. These convert information coming in (from a given set of sensory organs) into actions (realised by a given set of motor organs). In the simplest case we regard this as a one-stage procedure. There is a data input y which is converted into an action output u by a rule $u = \phi(y)$. The function ϕ is realised by the neural network and represents the response of the system. One would like to design the network so that this response is the appropriate one, on some criterion.

This response might be the response of a dog to a scent, of a factory manager to an order book or of an investor to market information.

By restricting attention to the one-stage case we of course exclude the fascinating and much wider issues raised by a changing environment, by the possibility that one's actions might affect that environment, or by the possibility of interaction with similar entities. These are indeed higher-order questions, whose resolution must rest upon that of the simplest case.

The response function ϕ is termed a 'decision function' or an 'action rule' in other contexts, and statistical decision theory gives a basis for the optimisation of ϕ. However, this optimisation requires the specification of a full statistical model and optimisation criterion. Furthermore, it gives no

3

guide to the physical realisation of ϕ and takes no account of the cost of such realisation. It presumably gives an idea of ideal performance, although the optimal solution may be non-robust. That is, performance of the rule deduced may degenerate disastrously if conditions depart even slightly from those assumed. The only protection against this is to optimise average performance over a sufficiently wide class of models.

The essence of the artificial neural net (ANN) approach is that the response rule is improved by evolution under actual operation. The neural net is modified (usually by perturbation of parameters) through the application of a 'learning rule' which improves performance in the light of experience. This offers the prospect of a procedure which is self-optimising under a wide range of assumptions, and so is adaptable, robust and self-righting. It might also give insight into the emergence of structure, and into the functioning of biological analogues. It constitutes the 'ultimate package', in being to a high degree assumption- and analysis-free. However, some ingenuity must be built into the learning rule if this is to be reasonably economical in time and effort. The mutation/selection rules of Darwinian evolution are plainly effective, but levy a vast cost in time and organic material.

However, to characterise operation in terms of a blind response rule is too shallow, even for the simple one-stage case, and this is implicitly recognised in ANN theory. The data y may be regarded as an imperfect indicator of the actual 'state' x of (he environment ('state' at least in relevant respects). So, the state for the dog above would be a complete identification of the cause of the scent. If one knew what x was, then decision would be easier to the extent that uncertainty would have been eliminated. Under at least some circumstances, and at least approximately, the deduction of an optimal action can be decomposed into the formation of an 'estimate' $\hat{x}$ of x from y, followed by the determination of the action u appropriate to $\hat{x}$.

The perception of the concepts of state and its estimation could be regarded as the first glimmer of intelligence on the part of the optimiser, although perhaps not yet on the part of the net itself. A device which simply forms this estimate $\hat{x}$ is an *associative memory*, in that it evokes the value $\hat{x}$ of x which is most consistent with the observations y, in some sense. A device which further holds this 'memory trace' in evoked form over time is a *storage memory*. So, time can be relevant, even for a one-stage system. In fact, dynamic effects can enter naturally in quite another way. In seeking the x which is 'most consistent' with observations y we are looking for a value of x which is extremal in some sense. This extremal problem is best solved, not by a pedestrian search through x values, but by a system of dynamic equations which seeks the extremum much more immediately — with much more sense of the geometry of x-space. Beckerman (1997) gives a graphic example of this: a protein model, governed by high-dimensional nonlinear dynamics, finds its minimal-energy configuration in about 10^{-10} of the time that an exhaustive search would take.

In fact, these two reasons for dynamic operation combine in a curious way, which finds expression first in Chapter 8. The function of a full associative memory is to form an inference on the basis of data which may be fleeting and which cannot be stored without degradation — what we shall henceforth refer to as 'fading data'. The dynamics must then be such as to combine the operations of inference (i.e. of search for the best-matching memory trace) and of retention of the significant aspects of the data during this search. In the end, the only aspect of data retained is identification of the best match.

We shall deliberately restrict ourselves to the estimation problem; if one can deal with it, one can probably deal also with the action problem. Even this reduced task is substantial enough; there is of course an enormous literature on associative memories. However, much of this is concerned with the discussion of relatively *ad hoc* proposals, and attention is split between the studies of performance of the net and of performance of learning rules.

We shall also restrict ourselves purely to the design problem. So, central as the learning aspect is to ANN theory, we shall not consider it (save in the survey of Chapter 10) but shall return to classical statistical optimisation principles applied to an assumed model. However, these principles must be relaxed and given dynamic form if the rules deduced are to find natural realisation in a neural net, and also if they are to be able to deal with fading data.

The search for a design principle implies a concern for performance. This concern raises a number of points, one of them being that of compound memory traces; see Chapter 6. It also leads to an attempt to throw yet more light on that already highly exposed object: the Hopfield net. As we argue in Chapter 6, the net is plainly not 'right'; for one thing, it can deal only with memory traces of a very specific statistical character. Nevertheless, it exerts a fascination; a fascination which we assert stems from the fact that the net indeed addresses the fading data problem, even if not very explicitly intended to do so.

The fading data problem is introduced in Chapter 8, and resolved by appeal to a probability-maximising principle. The probability-maximising algorithm (PMA) thus derived is almost trivially simple: in effect, a dynamic, autoassociative and somewhat fuzzy form of the Hamming net treated in Chapter 5. It thus has an immediate network implementation, and we show (Chapter 9) that it has optimality properties. Simplicity does not imply triviality; the Hamming net has structure which develops unexpectedly when some degree of compounding of traces is permitted; even more when oscillatory operation is added.

The frequency of reference to biological models and biological reality increases as we progress through the text. This is partly because these constitute what John Taylor, among others, has referred to as the only existence proof we have (of neural nets which function at the level of performance that one would wish). It is then natural to seek guidance from them. Nature demonstrates that neural networks successful in every relevant sense (scale, performance, reliability, adaptability, economy) really are possible, and can indeed show a sophistication infinitely beyond present ANN aspirations.

At a much more specific level, the mathematics seem to indicate that reliable performance at positive information rates can be reconciled with economy of realisation only if one considers compound stimuli (Chapter 6). We are then led to consider the example of the olfactory system, widely studied because of its relative simplicity. This is a good example, in that the odours activating it can plainly be simply superimposed, so constituting additively-compounded stimuli. The visual system is, of course, studied even more, but is greatly more complicated in that it has to cope with spatial structure, non-additive compounding, movement, and the recognition of images subject, not merely to superficial distortion, but to whole families of radical transformations.

The example of the olfactory system is considered explicitly in Chapter 14, where we argue that the net derived from the PMA provides a very plausible model of the transmission/processing loop constituted by the olfactory bulb and the anterior olfactory nucleus. The literature is of course littered with wishful, facile and deluded claims of this type. However, there are striking and unforced correspondences between the two structures on a number of significant features. Indeed, we are led to make testable predictions on the strength of them.

In Part III we follow up the commonplace of observation: that most biological neural nets (BNNs) are oscillatory in their functioning. The obvious reason for such a mode of operation is that suggested in the preface: that absolute activity levels are almost meaningless in the biological context. Information must then be transmitted by a variable rather than a static signal. An oscillatory signal can then be seen as a way of transmitting information despite a background of both noise and an irrelevant slow variation in baseline activity.

Quite the most impressive and thorough investigation of this matter has been that sustained by W.J. Freeman over a number of years (see the references under his name, plus the other references quoted in Chapter 11). Freeman has particularly studied the olfactory system, at the levels on the one hand of anatomy and experiment, and on the other of modelling, mathematical analysis and analogue simulation. He has developed a model of neural masses which retains essential neuronal dynamics, has derived a physiologically faithful model of a fundamental oscillator which has quite specific properties as a threshold element, and has demonstrated that systems can be built up from these components which show the character and behaviour of living systems.

Part III is devoted to development of Freeman's proposals. We set out his general ideas in Chapter 11 and develop some of his models in Chapter 12. However, as mentioned in the preface, the detailed structures Freeman suggests for the associative-memory function do not go beyond the Hopfield net. In Section 12.8 and Chapter 13 we transfer the PMA of Chapter 8 to the Freeman context; some interesting effects emerge. The consequent net is compared with olfactory anatomy in Chapter 14.

Use of the term 'chaotic' in the title may raise associations or expectations which we do not intend. We use it as Freeman does: not as characterised by Hausdorff dimension or by an infinitely fine-grained sensitivity to initial conditions (meaningless in this context!), but by a pattern of oscillation which is complex to the point of apparent irregularity. This complexity is an indication of robustness rather than sensitivity. When one sub-system passes signals to another (e.g. the olfactory bulb to the pyriform cortex) it is not a question of the second sub-system's 'resonating' to a pure carrier frequency of the first. The two constitute components of a joint system, coupled in both directions, and the oscillations are those characteristic of the joint system. Hence both their complexity and yet their coherence between sub-systems. The oscillation is a 'carrier' principally in that its relative amplitude in different parts of the cortex (say) is characteristic of the external stimulus applied to the system. More exactly, it is characteristic of the class (of a defined set of classes) to which that stimulus belongs.

Both actual organisms and Freeman's electronic analogue show oscillations in neural potential of a character which one would regard as chaotic. This is to be expected; a consequence of the necessary complexity and nonlinearity of these systems and of the inevitable transmission delays. The point is that the systems can be seen as functioning in spite of chaos, rather than in virtue of it.

Organisation of the text

We have already used the only three acronyms which we shall find habit-forming: ANN (artificial neural net), BNN (biological neural net) and PMA (probability-maximising algorithm). In Chapter 14 a number of standard anatomical acronyms are used, and are explained there. Vectors are column vectors, unless otherwise stated.

Sections, equations, theorems and figures are numbered consecutively throughout a chapter, with a chapter prefix. So, equation (3.18) is equation (18) of Chapter 3. However, the chapter prefix is often omitted when reference is made within a chapter.

Chapter 2

Optimal Statistical Procedures

The ANN approach to the evolution of good action rules is, superficially, quite different to that of statistical decision theory, and, indeed, a reaction away from it. Statistical decision theory seeks procedures which are optimal under well-defined conditions and criteria; in the ANN approach one seeks procedures which are robust in that they perform well under a range of conditions. Statistical decision theory supposes enough information available that realistic models can be formulated, whose rational analysis gives a basis for action. The ANN approach seeks 'merely' a simple improvement rule whose continued application will lead, in the course of time, to action rules which are appropriate under the conditions experienced. Statistical decision theory takes no explicit account of the computational burden of implementation; in the ANN approach a strict discipline is enforced by the requirement that all calculations should be realised within the network itself.

Nevertheless, the search for optimality provides an invaluable guide to what is good and what is significant, and any procedure derived as 'optimal' can always be tested for robustness and realisability. One might also note that an increasing proportion of ANN work resembles and overlaps the traditional statistical approach, which is the reason for the qualification 'superficially' above.

Statistics is a large subject, and we shall cover only what is immediately relevant in as simple a fashion as possible. Distinct from the analysis of models is what one might term 'canonical reduction', which we touch on to some extent in Chapter 10.

2.1. The optimisation of actions

The basic formalised problem is that one must take an action u on the basis of information y. For example, an animal might catch a scent, and

will immediately try to assess it, with the possibilities of food, sex, danger
or threat to territory in mind, and the immediate possible actions of
further investigation, flight, pursuit (cautious or aggressive) and so on. We
suppose the decision to be an isolated one, but, as this example shows,
in actual life information and decision follow each other in a chain of
chance and implication, and a more adequate theory treats the problem as a
dynamic one.

Suppose that past experience has taught that, if one takes action u in
situation y, then one experiences a cost C whose average value (in these
circumstances) we can write as $E(C|y, u)$. That is, C and y are regarded as
joint random variables, and $E(C|y, u)$ is the expected value of C *conditional*
on the value y of the 'situation' and *parametrised* by the value u of the action
taken. It follows that, if minimisation of expected cost is one's goal, then
in situation y one should choose u to minimise $E(C|y, u)$.

However, this rule can be given a more meaningful form, which contains
the beginnings of a model. Suppose that the cost of taking action u can be
written as a function $C(x, u)$ of u and of x, where x represents 'the true
state of affairs' — what is often called the 'state of nature', or simply the
'state'. For example, in the case of the animal above trying to identify
a scent, x would correspond to such a complete identification — species,
numbers, ages, sexes, dispositions etc. of the animals (we assume) which
left the scent. If the animal knew x (and was a rational decision-taker) it
would know what to do: it would choose u to minimise $C(x, u)$. However,
it knows only y, which one can regard as a random variable associated
with x. In fact, let us regard x and y as joint random variables with a joint
distribution $P(x, y)$. (For simplicity of presentation, we shall for the moment
assume variables discrete-valued. This constitutes no real loss of generality;
we explain our notation and indicate the formal passage to the general
case in Note 1.) We can write $P(x, y) = P(x)P(y|x)$. The conditional
distribution $P(y|x)$, the distribution of observation conditional on state,
would reasonably be specified in a model. The distribution $P(x)$ is termed
the *prior distribution* of state; it expresses one's assessment of the fraction
of cases in which state would take the value x in a particular case if one had
no information on that case. So, in the case of the animal coming upon a
scent, $P(x)$ would express the probabilities of the different contingencies
which might yield a scent, as assessed *before* this particular scent was
detected. For example, an urban dog can well expect to find traces of
other dogs or (less interestingly) of humans, but is unlikely to find evidence
of sheep.

The action can depend only upon the observable y, so a specific action rule must take the form of a function $\phi(y)$ from the space of situations y to the space of actions u. The expected cost under this rule would be

$$\sum_x \sum_y P(x,y)C(x,\phi(y)) = \sum_x \sum_y P(x)P(y|x)C(x,\phi(y))$$

$$\geq \sum_y \min_u \left[\sum_x P(x)P(y|x)C(x,u) \right] \qquad (2.1)$$

Theorem 2.1 (The optimal decision rule with imperfect information). *The optimal value of u for given y is that minimising the expression*

$$\sum_x P(x)P(y|x)C(x,u) \propto E[C(x,u)|y] \qquad (2.2)$$

Proof. The rule suggested gives equality in the inequality of (1), and is admissible in that it makes u a function of y alone. $\qquad \square$

The second expression in (2) is proportional to the first as a function of u, so the rule derived is consistent with that given earlier. The assertion of the theorem may seem self-evident, but its implication deserves emphasis: that, for this problem, minimisation with respect to a function can be reduced to minimisation with respect to a single value.

Notes

1. We use $P(x)$ to denote the probability that the *random variable* labelled by x takes the *particular value* x. There is thus a blurring of the distinction between a random variable and the values it may assume, but the notation is an uncluttered one which needs rescue from ambiguity only occasionally. If we wished to express the probability that the random variable x took a particular value a then we would write $P(x = a)$.

   One would generalise the treatment by, for example, replacing an expectation such as $\sum_x P(x)\phi(x)$ by $\int f(x)\phi(x)\mu(dx)$. Here μ is a measure and f a density with respect to that measure. A fixed measure will generally be adopted for a given treatment (e.g. counting measure for the discrete sums employed in the text) while the density f may be varied (as the distribution $P(x)$ could be varied in the text).

2. A special case of interest is that of inference on the value of state itself, so that the action $u = x_1$ means that one regards the value x_1 as being

the most plausible value of state x. This occurs in the communication context: x constitutes 'message intended' (which is assumed to take values in a finite set χ), y constitutes 'message received' (which may be a highly distorted version of x), and one has the task of deciding, on the basis of y, what x might have been. Suppose that the cost is zero if one makes the correct decision and is unity otherwise, so that $C(x, u) = 1 - \delta_{xu}$. Then the rule minimising the probability of error is: choose u as the value of x maximising $P(x)P(y|x)$.

2.2. Effective estimation of state

In many cases one can gather increasing amounts of data. This might, for example, be reflected in the fact that y is a vector whose dimension n, potentially variable, is a measure of the amount of data available. Under some commonly valid statistical assumptions this can imply an ever more accurate determination of the unknown state x, in that the term $P(y|x)$ occurring in the sum $\sum_x P(x)P(y|x)C(x, u)$ shows an ever sharper maximum around a value $\hat{x}$, which presumably is an estimate of the true value of x. The sum is then dominated by the term, for which $x = \hat{x}$ and the optimal u is approximately that value maximising $C(\hat{x}, u)$.

Here x appears effectively as a parameter of the conditioned y-distribution $P(y|x)$, and $\hat{x}$ is the *maximum likelihood estimate* of this parameter. Under regularity conditions the maximum likelihood estimate can be shown to be *consistent*, in that it converges stochastically (in several specific senses) to the true value of x as n increases. It can also be shown to be *asymptotically optimal*, in that it has asymptotically the smallest possible mean square deviation from x. (See e.g. Rice, 1995.)

What is interesting is that an attempt at action optimisation has generated an *estimate* $\hat{x}$ of x. Further, it has reduced the optimisation problem to what it would have been if x had the known value $\hat{x}$. This amounts to a statement of *certainty equivalence*: that one can optimise action as though the estimated value of state were the actual value.

What we have made, plausible is that certainty equivalence will hold asymptotically under regularity and convergence conditions. It will scarcely ever hold exactly, and there are situations whose essence is that it does *not* hold (see the note). However, in the next section we examine one important class of cases for which a certainty equivalence principle does indeed hold exactly.

Note

Suppose that an animal may spend time foraging locally at an assured rate of return (of nourishment) a per day, or hunting elsewhere with an unknown rate of return x per day. If it spends a fraction u of its day locally then the total return over the day is $au + x(1-u)$. If the animal knew the value of x and was rational then it would of course invest all its time in the activity of greater return. Suppose u is chosen to maximise $E[(au+x(1-u))^\nu|y]$, where y represents all the information available. If $\nu = 1$ (indicating *risk-neutrality* on the part of the animal) then the animal should spend all its time in the activity of greater expected return. That is, certainty equivalence holds, with $\hat{x} = E(x|y)$.

However, if $0 < \nu < 1$ (indicating *risk-averseness*), then the optimal u will in general lie strictly between 0 and 1. That is, the animal splits its time, which is inconsistent with certainty equivalence. This amounts to *hedging* in the equivalent investment context.

2.3. The quadratic/Gaussian case: Estimation and certainty equivalence

It made little difference in the last section whether we took $\hat{x}$ as the value of x maximising $P(y|x)$ or the value maximising $P(x,y)$, since the two expressions differ by a factor $P(x)$, independent of n. However, for the exact results of this section we must take the second definition.

Suppose that x and y are random vectors, reduced to zero mean, for convenience, and jointly normally distributed. The logarithm of their joint probability density $f(x,y)$ is then of the form const. $-\frac{1}{2}Q(x,y)$, where Q is a positive definite quadratic form in the two vectors. If $\hat{x}$ is the value of x maximising f (and so minimising Q), then this implies a decomposition

$$Q(x,y) = Q_1(x - \hat{x}) + Q_2(y), \tag{2.3}$$

where Q_1 and Q_2 are also positive definite quadratic forms in the variables indicated.

Decomposition (3) implies that the estimation error $\Delta = \hat{x} - x$ and the data vector y are normally and independently distributed, with zero means. This in turn implies the certainty equivalence property if the cost function $C(x, u)$ is quadratic. More explicitly:

Theorem 2.2. *Suppose the state variable x, the data y and the action variable u are all vector-valued. Suppose that x and y are jointly normally* *distributed, and let $\hat{x}$ be the value of x maximising their joint probability density. Then*

(i) *$\hat{x}$ can be identified with $E(x|y)$.*
(ii) *Suppose also that the cost function $C(x, u)$ is quadratic in both vectors, positive definite in u. Then the optimal value of u is that minimising $C(\hat{x}, u)$.*

Proof. The first assertion follows from the independence of Δ and y expressed by (3). From this independence and the assumptions on C it also follows that

$$E[C(x, u)|y] = C(\hat{x}, u) + \text{terms independent of } u. \qquad (2.4)$$

To see this, write $x = \hat{x} - \Delta$, and expand under the expectation in powers of Δ. The first-degree term has zero conditional expectation, in virtue of the independence asserted above. Because of the quadratic character of C, the second-degree term is independent of u and there are no terms of higher degree. Since the optimal value of u minimises the expectation in (4), the second assertion of the theorem (a statement of certainty equivalence) follows. □

The independence expressed by the decomposition (3) also implies that the elements of Δ are uncorrelated with those of y. We can write this as $\Delta \perp y$, a statement that the estimation error is statistically orthogonal to the data vector. This orthogonality property on its own, without the assumption of normality, guarantees certain optimality properties of the estimate. We summarise these, leaving proof to the reader (or any of a multitude of texts; Whittle (1996) despatches it).

Theorem 2.3. *Let u and y be random vectors, standardised to zero mean and possessing second-order moments. Let $\bar{x}$ be any linear function of y, so written because it is regarded as a possible estimate of y. Let $\hat{x}$ be any such linear estimate which also obeys the orthogonality condition $\hat{x} - x \perp y$. Then*

(i) *The orthogonality condition $\bar{x} - x \perp y$ is necessary and sufficient for the minimum mean-square error property: that in the class of linear estimators $\bar{x}$ should minimise $E(\bar{x} - x)^T G(\bar{x} - x)$. Here G is any positive definite matrix with the dimension of x.*
(ii) *Equivalently, $E(\bar{x} - x)(\bar{x} - x)^T \geq E(\hat{x} - x)(\hat{x} - x)^T$, in the sense that the difference of the two matrices is non-negative definite.*
(iii) *$\hat{x}$ is unique up to a mean-square equivalence.*

We shall term $\hat{x}$ the *projection estimate*, because of this minimum mean-square property. It is characterised by the orthogonality property, but has the two dual extremal characterisations (the *Gauss-Markov theorem*). One is that $\hat{x}$ is the value of x minimising the quadratic form $Q(x, y)$ appearing in the Gaussian exponent. The other is that it can be written By, where B is chosen to minimise the covariance matrix of $x - By$ (in the positive definite sense). So, in the one case the extremisation is with respect to the variable itself; in the other with respect to the linear operation defining the variable.

The second characterisation yields the well-known explicit form

$$\hat{x} = V_{xy} V_{yy}^{-1} y \tag{2.5}$$

for $\hat{x}$, where V_{xy} is the covariance matrix $E(xy^T)$. This expression is meaningful even if V_{yy} is singular, which is just the case for which assertion (iii) of the theorem has some point.

2.4. The linear model, in Bayesian and classic versions

Consider the case when the data vector y is assumed to be related to the state vector x by

$$y = Ax + \epsilon \tag{2.6}$$

Here A is a known matrix and ϵ is a vector of observation errors, supposed random and independent of x. The data vector is thus a linear function of the state vector, plus observation error.

Let us suppose that x and ϵ are normally and independently distributed, with zero means (achievable by replacement of x by $x - E(x)$, etc.) and covariance matrices U and V respectively. With this the joint distribution of x and y is determined, and we find that the quadratic form $Q(x, y)$ in the exponent of their joint Gaussian distribution is

$$Q(x, y) = x^T U^{-1} x + (y - Ax)^T V^{-1}(y - Ax). \tag{2.7}$$

The projection estimate $\hat{x}$ minimises this expression, and so satisfies the linear equation

$$(U^{-1} + A^T V^{-1} A)\hat{x} = A^T V^{-1} y. \tag{2.8}$$

Solutions (5) and (8) for $\hat{x}$ necessarily agree — a fact which implies some interesting matrix identities.

Model (6) has been central in statistics as 'the linear model', traditionally regarded as expressing the observation vector y linearly in terms of the columns a_j of A, rather than linearly in terms of x. The elements x_j of x then, appear as parameters of the model, the *regression coefficients*, and are not classically given a prior distribution. We can recover the classic treatment formally by letting U tend to infinity, which is a way of saying that there is no useful prior information. Very often it is also assumed that the individual errors ϵ_k are independently and identically distributed, so that V is proportional to the identity. With these simplifications (8) takes the form

$$A^T A \hat{x} = A^T y, \tag{2.9}$$

the classic equation for the estimate $\hat{x}$ of the vector of regression coefficients. This is classically termed the *least square estimate*, not in the sense that it itself has minimum mean-square error (a property characterised as 'minimum variance unbiased' in this case), but in the dual sense: that it minimises the quadratic form to which $Q(x, y)$ reduces in this case: the 'residual sum of squares' $|\epsilon|^2 = |y - Ax|^2$.

Chapter 3

Linear Links and Nonlinear Knots:
The Basic Neural Net

In common with natural neural networks, ANNs are synthesised from virtually one basic component: an analogue of the neuron. The simplest such analogue is the linear gate, which is in fact the sole algorithmic component of the majority of ANNs. Associated with this is the particular nonlinearity displayed by neurons: the sigmoid response.

3.1. Neural calculations: The linear gate and the McCulloch–Pitts net

Procedures derived from the statistical criteria of Chapter 2 set the limit for what is achievable in the way of performance under the criterion assumed if: (1) reality indeed follows the model assumed, and (2) there is access to unlimited and costless computing facilities. ANN procedures may seem to be so assumptionless as to be shapeless, but they are in fact strongly optimality-seeking: the 'optimality' which is achieved in nature by the mutation/selection of Darwinian evolution. Moreover, this concept of optimality respects two harsh constraints. One is that of *robustness*: that methods must work well under a range of conditions. The other is that of *self-sufficiency*: that calculations must be realised within the network itself. Optimality trade-offs will then force the realisation to be an economical one.

The great majority of artificial neural nets are made up of a single type of component, realising a single type of operation. This is the *linear gate*, regarded as the simplest nontrivial analogue of the biological neuron. (Strictly speaking, it is the analogue of a *neural mass*; a point to which we return in Chapter 11.) Suppose that a node of the net has scalar inputs

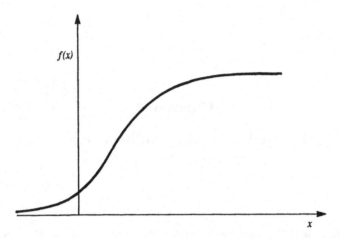

Fig. 3.1.   A sigmoid function; in fact, the generic soft threshold function.

$y_1, y_2, \ldots$ from other nodes, constituting an input vector y, and that it yields a scalar output η. Then one assumes the relation

$$\eta = f\left(\sum_k w_k y_k\right) = f(w^T y).\tag{3.1}$$

Here the w_k are constant weighting coefficients (*synaptic weights*), constituting a vector w, and f is an *activation function* with the sigmoid form of Fig. 3.1. We shall define the sigmoidal property more exactly in the next section.

The linear gate thus embodies two operations: the linear combination of inputs followed by a nonlinear response of the sigmoid form. It is this sigmoid response which provides the essential nonlinear element of a neural calculus. We shall regard the gate as a basic unit, the simplest artificial neuron. It seems to embody the most obvious features of the natural neuron, and these features in fact give it a high degree of computational versatility. Of course, the natural neuron is much more complicated; in particular, it has much more in the way of internal dynamics. We return to this point in Section 4 and, more extensively, in Chapter 11.

We can build these neurons into a net of n nodes and achieve some degree of dynamic operation by writing the relation holding at the jth node as

$$y_j(t+1) = f\left[\sum_k w_{jk} y_k(t)\right], \quad (j = 1, 2, \ldots, n; t = 0, 1, 2, \ldots),\tag{3.2}$$

where $y_j(t)$ is the output from node j at time t. That is, the net operates in discrete time, with time advancing in unit steps. Each node operates as a linear gate, but with a unit delay.

The y_j are assumed to be scalar-valued. They are the analogues of the firing rates of natural neurons (or, more precisely, pulse rates of neural masses). The firing rate interpretation would constrain them to be non-negative; the synaptic weights w_{jk} can on the other hand be positive or negative according as they represent an 'excitatory' or an 'inhibitory' input to the jth neuron. However, in the ANN context the activations are permitted to take either sign; non-negativity can always be restored by a linear transformation of the variable.

Nets of the type (2) were proposed first in a seminal paper by McCulloch and Pitts (1943), for which reason they are called McCulloch–Pitts nets. Actually, these authors took the extreme form of activation function $f(\xi) = \mathrm{sgn}(\xi)$, implying that the y_j take only the values ± 1 (see the note). A net of this nature is often termed *binary*. The term 'linear gate' would often be understood to imply a signum activation function. However, we shall understand the linear gate to have the action (1) with a general activation function, and shall term the signum function a *hard threshold function* (see Fig. 3.2), in contrast to the *soft threshold function* represented in Fig. 3.1 — definitions to be formalised in the next section.

The relations (2) of the McCulloch–Pitts net can be written more compactly in the vector form

$$y(t+1) = f[Wy(t)]. \tag{3.3}$$

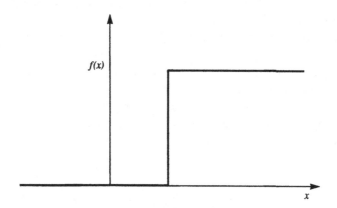

Fig. 3.2.   A hard threshold function.

Here W is the *synoptic matrix*, with elements w_{jk}, and we have taken the convention that f applies to vectors componentwise. That is, $f(x)$ is the vector with elements $f(x_j)$.

Grossberg (1988) includes an excellent short historical review of early models of BNNs, equally valid as early proposals for ANNs. He takes exception to authors who refer to the McCulloch–Pitts net (3) as a 'Hopfield net'. The Hopfield net is a particular case of such a net, and, indeed, one with a pre-Hopfield history; see Nakano (1972), Kohonen (1972), Anderson (1972) and Amari (1972, 1977a).

An input can be fed into net (3) either by specification of the initial value, $y(0) = y$, say, or as a sustained driving signal u:

$$y(t + 1) = f[Wy(t) + u(t)]. \tag{3.4}$$

The question is whether an external stimulus is momentary (but vigorous enough to reset the whole net) or sustained and with a cumulative effect. The two cases differ crucially, and we shall consider both.

The so-called *feedforward* nets constitute an important subclass which should be defined, even if it is less relevant in our context. In these nets the nodes are divided into ordered *layers*, and propagation is unidirectional in that synaptic connections are possible only to a higher layer. That is, w_{jk} can be non-zero only if node j lies in a higher layer than does node k. The first (lowest) layer is taken as the *input layer*, and activations here take prescribed input values. The last layer is correspondingly the *output layer*. Intermediate layers are termed the *hidden layers*, since they have no direct connection either way with the outside world. If there are r layers ($r \geq 2$) with connections only between adjacent layers then the system will deliver the output for a given input after a delay of $r - 1$ units of time. The input/output relation is determined by the synaptic matrix W; the layering constraint still allows plenty of latitude in its choice. It can be shown that, if the net is built of hard-threshold linear gates, then any Boolean input/output function can be realised (indeed, with a single hidden layer), and that, with soft-threshold gates, virtually any function of interest can be approximated arbitrarily well. Hassoun (1995) gives an excellent review of these matters.

Note

The signum function $\mathrm{sgn}(x)$ takes the values $+1$ or -1 according as x is positive or negative, but conventions differ as to what value it is assigned for zero x. Sometimes it is given the value zero, sometimes one of the values

±1 as an arbitrary resolution of the tie. In most of our discussions the argument x is a continuous variable, so the point virtually never arises. See Theorem 3.1 of Section 3.

3.2. Sigmoid and threshold functions

The loosest characterisation of a sigmoid function $f(x)$ is that it is a non-decreasing function of a scalar variable x, bounded above and below. It is thus very much like a probability distribution function, and could be normalised to one by a linear transformation (so that $f(x)$ increases from 0 to 1 as x increases from $-\infty$ to $+\infty$) if we exclude the trivial case of constant f. Indeed, in one treatment of neural dynamics (see Section 11.2) $f(x)$ represents just the proportion of neurons in a population which would 'fire' if they received an input of x. That is, the proportion of the population whose firing threshold does not exceed x.

A more specific class, and one corresponding more closely to the usual intuitive picture, is what we shall term the class of *threshold functions*, denoted $\mathcal{T}$. These are the sigmoid functions $f(x)$ which are convex in some interval $x \leq x^*$ and concave in $x \geq x^*$. The value x^* can be regarded as a threshold in that it separates regions of what one might regard as 'sinking' and 'rising' inclination. The 'distribution' corresponding to a threshold function is *unimodal*, in that its density $f'(x)$ (if this exists) is non-decreasing for $x \leq x^*$ and non-increasing for $x > x^*$ (and, of course, has a finite integral).

If a threshold function has a jump, corresponding to a δ-function component in the density, then this can occur only at x^*. The extreme case of this would be the *hard threshold function*, for which $f(x)$ increases only at x^*. We shall take a *soft threshold function* as being a threshold function for which $f'(x)$ exists (and is then necessarily unimodal). Under appropriate continuity assumptions we can assert that the maximal value of the derivative must occur at x^*, which will also mark the sole point of inflection of $f(x)$.

The real roots of the equation $f(x) = x$ are important; these are the *fixed points* of the transformation f. The convex/concave character of a threshold function implies that there can be at most three such roots, as in Fig. 3.3. A shift of f could reduce these to one.

Note

One important threshold function is the *logistic function* $f(x) = (1+e^{-\beta x})^{-1}$, with the attractive property that $f'(x) = \beta f(1 - f)$. This is sometimes taken

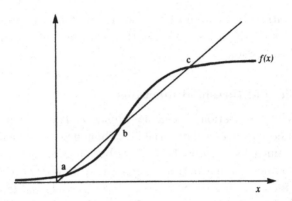

Fig. 3.3.   An illustration of the three fixed points of the sigmoid $f(x)$.

as an expectation version of a stochastic model, in which the output variable y can take the values 0 or 1 with respective probabilities $1 - f(x)$ and $f(x)$ if the input is x. Otherwise expressed: $P(y = 1|x)/P(y = 0|x) = e^{\beta x}$.

3.3. Iteration

If the threshold function f in (3) is made softer, then in general the number of equilibrium solutions of the recursion decreases, but the basins of attraction of these equilibria widen. That is, if the network serves the purpose of 'classifying' the initial input $y(0) = y$ by the equilibrium to which the recursion (3) then converges, then a softening of the threshold function leads to a coarser classification, but one more tolerant of variation in the input. A degree of softness may then be desirable, as we shall see in Chapter 8.

However, the mere fact of convergence (if it occurs, which has yet to be demonstrated) implies that iteration of a soft threshold function can produce the effect of a hard one. This was the point made by von Neumann (1956) when he corrected error by a 'restoring organ', realised by iteration of an appropriate operator. Let us examine the effect of simple iteration of f.

Let the tth iterate of $f(x)$ be denoted $f(x,t)$, so that $f(x,1) = f(x)$ and the recursion $f(x,t+1) = f(f(x,t))$ holds for $t = 0,1,2,\ldots$. This then implies the seemingly more general identity

$$f(x, s + t) = f(f(x,t), s) \quad (s, t = 0, 1, 2, \ldots), \tag{3.5}$$

self-consistent if we set $f(x, 0) = x$.

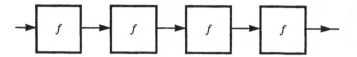

Fig. 3.4.   A sequence of operations, realising iteration of $f(x)$.

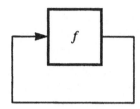

Fig. 3.5.   Realisation of iteration by feedback.

One could view this iteration as being realised by a sequence of operations, corresponding to the blocks of Fig. 3.4. However, the more natural realisation is by *feedback*, Fig. 3.5, which replaces replication in hardware by repetition in time. Expressed formally, if we consider the recursion

$$x(t + 1) = f(x(t)), \quad (t = 0, 1, 2, \ldots) \tag{3.6}$$

then the solution in terms of the initial value $x(0) = x$ is just $x(t) = f(x, t)$.

Theorem 3.1. *Suppose that the soft threshold function $f(x)$ is such that it has the full set of three fixed points, at $x = a, b, c$, say $(a < b < c)$. Then all iterates $f(x, t)$ have these same fixed points. As t increases $f(x, t)$ converges to the hard threshold function*

$$f(x, \infty) = \begin{matrix} a & (x < b) \\ b & (x = b) \\ c & (x > b) \end{matrix} \tag{3.7}$$

The convergence is monotone in that $f(x, t+1) - f(x, t)$ has the same sign for $t = 0, 1, 2, \ldots$ for any given x.

Proof.   The proof follows from the feedback characterisation (6) with $x(0) = x$. The point $x = b$ is an equilibrium of the recursion; if $x < b$ then $x(t)$ converges monotonically to a; if $x > b$ then $x(t)$ converges monotonically to c. This implies all the assertions of the proposition. $\square$

The convergence of $f(x,t)$ with increasing t to the hard threshold function is thus just a consequence of the fact that recursion (6) will seek one or other of its two stable equilibria: a or c. To start at the unstable equilibrium point b and stay there is in practice an event of probability zero. The effect of the sigmoid element in a network is expressed most succintly just by this bistable character of the threshold/feedback unit. It is this which produces the 'quantisation' in more general networks exemplified by a multiplicity of distinct equilibria and by the suppression of minor features in favour of major ones.

Theorem 3.1, with its associated identification of the concepts of iteration and of feedback, has an immediate continuous-time analogue. The iterative property is carried over by requiring relation (5) to hold for all s, $t \geq 0$. In the discrete-time case the relation $f(x,0) = x$ held trivially, and we had to start the iteration by specifying $f(x,1)$ as the given $f(x)$. In the continuous case we have to specify an 'infinitesimal generator'; i.e. suppose that $f(x,t) = x + tg(x) + o(t)$ for some $g(x)$ for small positive t. It is the perturbing $tg(x)$ which marks the deviation from linearity and gives $f(x,t)$ its sigmoid character. If we demand the threshold property with a full set of three fixed points, then $g(x)$ must have the form illustrated in Fig. 3.6.

Corresponding to the assertion associated with (7), we can row regard $f(x,t)$ as the solution $x(t)$ of the differential equation

$$\dot{x} = g(x) \tag{3.8}$$

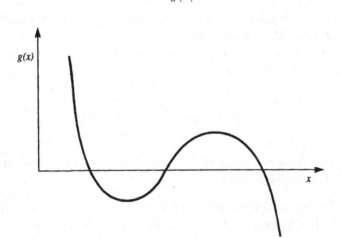

Fig. 3.6.   The infinitesimal generator $g(x)$ of a continuous family of sigmoid functions.

for prescribed initial value $x(0) = x$. The assertions of Theorem 1 plainly transfer to this continuous-time case.

In fact, we can be more explicit in the continuous-time case. Equation (8) can be immediately integrated to yield the relation

$$\int_x^{f(x,t)} \frac{d\xi}{g(\xi)} = t,$$

where x and $f(x,t)$ necessarily lie on the same side of b.

Notes

1. Suppose we write (8) as $\dot{x} = -\partial V(x)/\partial x$, so that the dynamics seek a minimum of the 'potential function' $V(x)$. If g has the form illustrated in Fig. 3.6 then $V(x)$ has the form illustrated in Fig. 3.7, with turning points at a, b and c. Convergence of the solution of (8) to one of the stable equilibria $x = a$ or $x = c$ corresponds to capture in one of the two potential wells.

2. Passage to the limit in the differential equation (8) achieved a hard threshold response to initial value. It can achieve at least a hardened threshold response to a sustained constant input u. Suppose that we have the driven equation

$$\dot{x} = g(x) + u \tag{3.9}$$

and that g has the form sketched in Fig. 3.8; special in that its point of inflection is also a point of zero gradient. This can be attained by

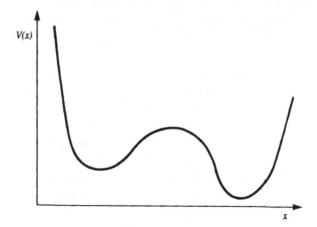

Fig. 3.7.   The double-welled potential from which $g(x)$ is derived.

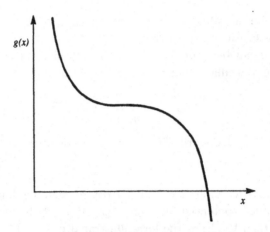

Fig. 3.8. The 'sensitive' infinitesimal generator.

subtracting a linear function from a sigmoid. Then the equilibrium solution of (9) is $x = g^{-1}(-u)$, which is a sigmoid with infinite gradient at its inflection point, and so a strong threshold response at that point. The equivalent $V(x)$ has a single 'fourth-power' potential well, which is skewed strongly if a linear biasing term ux is imposed. See Whittle (1991b).

3.4. Neural systems and feedback in continuous time

The linear gate is the most rudimentary possible model of the biological neuron. What is principally missing is anything in the way of internal dynamics. These must be introduced, not merely for biological faithfulness, but to allow an enrichment of performance which is certainly valuable and may be essential.

Suppose that the neuron has input u and output y. These are both regarded as scalar functions of time, separate inputs having been aggregated linearly to form the net input u. The linear gate then corresponds to the instantaneous relation $y = f(u)$, where f is a sigmoidal activation function.

A more adequate model introduces an intermediate variable x, the *membrane potential*. (We shall give both some anatomical detail and closer mathematical argument in Chapter 11.) The potential responds to input in a stable linear fashion, in that it obeys a constant-coefficient linear differential equation driven by input:

$$\mathcal{L}x = u. \tag{3.10}$$

The operator $\mathcal{L}$ can then be written as a polynomial $L(\mathcal{D})$ in the time-differential operator $\mathcal{D} = d/dt$. Stability requires that the zeros of the polynomial $L(s)$ should all have strictly negative real part.

The output variable y is regarded as *a firing rate*, and is related to the membrane potential by the simple sigmoidal relation

$$y = f(x). \tag{3.11}$$

This is the model particularly considered by Freeman (1975, 1987), who argues from observations on actual neurons that it is sufficient to take $\mathcal{L}$ as a second-order operator. Moreover, he also asserts that it cannot be of lower order if the effects of delay in the synaptic connections and of the charging dynamics of the membrane are to be represented. These facts turn out to be significant.

Return now to equation (8). A special case of this would be

$$\dot{x} = -kx + wf(x) \tag{3.12}$$

where k and w are positive constants and $f(x)$ is a threshold function. The right-hand member of (12) would indeed have the character demanded of g in Fig. 3.6 if the ratio w/k fell between appropriate bounds. But what equation (12) represents is a neuron of the type just described; particular in that $L(s) = s + k$, and that the only input is that supplied by the feedback of a proportion w of its own output.

We see that it is just this feedback which leads the system described by (12) to have three possible equilibria: a 'high-excitation' state, a 'low-excitation' state and an intermediate unstable state. The same possibilities hold for the more general system

$$\mathcal{L}x = wf(x), \tag{3.13}$$

k now being identified with $L(0)$; (see the note). Indeed, we can consider the vector version

$$\mathcal{L}x = Wf(x) \tag{3.14}$$

where x is now the N-vector of potentials of the N neurons and W is the *synaptic matrix* of connection weights between neurons. These connections of course represent feedback within the N-neuron system.

System (14) can be regarded as a dynamic version of the McCulloch–Pitts net, and the McCulloch–Pitts net as the equilibrium version of (14).

Suppose for simplicity that $L(0) = 1$, as can be achieved by a rescaling. The equilibrium version of (14) is then $x = Wf(x)$, which becomes

$$y = f(Wy) \qquad (3.15)$$

in terms of the vector of firing rates $y = f(x)$. This is just the equilibrium version of (3).

Note

The stable equilibria of system (12) are just at those zeros of $g(x) = -kx + wf(x)$ for which g has negative gradient. The stability criterion for system (14) is deduced in Theorem 12.1; for the scalar case (13) it reduces to the assertion that an equilibrium point $\bar{x}$ is stable if and only if $wf'(\bar{x}) < L(0)$.

3.5. Equilibrium excitation patterns

The continuous-time formulation of Section 4 is certainly the natural one physically, and the more general dynamical formulation (10), (11) is at least a more adequate representation of biological reality. Let us nevertheless return to the McCulloch–Pitts recursion (3), which has served for so many ANN models. If we work in terms of the 'potential' variable x related to y by $y = f(x)$ then the recursion becomes rather

$$x(t+1) = Wf(x). \qquad (3.16)$$

The natural expression of the fuller neural model is equation (14), an equation in the potential variable x, and (16) is the McCulloch–Pitts analogue.

We know that in the scalar case, $N = 1$, equation (16) has just two stable equilibria if the sigmoid has the right location and scale. The vector version would thus have 2^N stable equilibria if W were the right multiple of the identity. A question of great interest is then whether exponentially many (in N) stable equilibria exist for a given prescription of W as a function of N. This is a point which we pursue in Chapters 6 and 9, in particular.

3.6. Some special-purpose nets

Grossberg (1973, 1976a, 1976b, 1988) has proposed and analysed a number of particular dynamic systems, equivalent to nets, which effect special transformations as part of a larger purpose. One is the system

$$\dot{x}_j = -Ax_j + (B - x_j)y_j - x_j \sum_{k \neq j} y_k \qquad (3.17)$$

where j and k take the values $1, 2, \ldots, n$, and $y = (y_j)$ is an input vector, constant in time. This has the equilibrium solution

$$x_j = \frac{By_j}{A + \sum_k y_k},$$

so the transformation achieved is a scale-standardisation of the input vector y. This is something one would often desire, because the input vector can then be judged on its form, irrespective of its scale.

The combination $\dot{x} + Ax$ could be realised neurally as $\mathcal{L}x$, but the remaining terms in (17) are not reconcilable with the structure (14). The final sum represents a global inhibitory effect between nodes. Its coefficient x_j and the coefficient $(B - x_j)$ of y_j represent what Grossberg terms a 'shunting' mechanism, which has the effect of constraining x_j to lie between the bounds 0 and B. This is a mechanism by which he sets great store; it achieves effects which the purely additive feedback of (14) cannot. We shall find the need for such a feature, but shall see that the mathematics suggest a rather more 'neural' mechanism. Equivalent effects are achieved in nature by regulatory mechanisms which affect neuronal characteristics, usually on a slower time scale.

Another operation that one would like to realise is the identification of the maximal element of the input vector y. Networks achieving this are often termed 'winner-takes-all' nets.

One's first inclination is to carry out some kind of direct search, by direct comparison of elements of y. Lippmann (1987) describes the 'comparator' which does just that; it uses $O(\log_2 n)$ nodes, where n is the dimension of y.

One does much better if one constructs a dynamic system which by its nature seeks the maximum directly. Hassoun (1995) describes a number of such devices, based on mutual inhibition between nodes, and so on 'competition'. They all follow the same principle, which is most simply expressed in a continuous-time version. Suppose that the activations x_j obey the differential equations

$$\dot{x}_j = -\left[\sum_{k \neq j} x_k \right]_+ = -\left[\sum_k x_k - x_j \right]_+$$

with initial values $x_j(0) = y_j$. Here $[x]_+ = \max(0, x)$. Then, as long as there is more than one non-zero element, the non-zero elements will continue to decrease, preserving their ordering. As each element reaches the value zero it sticks there. The process continues until there is only one element left, which is that corresponding to the initially largest component of y.

The function $[x]_+$ shows sigmoid behaviour in that it saturates to a zero value for non-positive arguments. It is this feature which ensures that the irrelevant elements progressively drop out of the game. There is no call for saturation at large values of the argument, since all excitations are decreasing.

However, the most elegant way of all for solving an extremal problem, and possibly that used in nature, is to find a dynamical system which, by its nature, drops into a mode, static or dynamic, which is characteristic of the solution. As a rather general version of such a system, consider the set of equations

$$\dot{x}_j = g(x_j, z), \quad z = \sum_k f(x_k), \tag{3.18}$$

where $f(\xi)$ is monotonic increasing in ξ and $g(\xi, z)$ monotonic decreasing in z, and all differentials needed are supposed to exist. Equilibrium values of x_j must thus be roots of the equation

$$g(x_j, z) = 0, \tag{3.19}$$

with z given in terms of the x_k by the second relation of (18). Let us denote the values of $\partial g(x_j, z)/\partial x_j$, $\partial g(x_j, z)/\partial z$ and $df(x_j)/dx_j$ at the particular equilibrium considered by α_j, $-\beta_j$ and γ_j respectively. These will not in general be distinct for all j, because the x_j are restricted to the set of ξ-roots of the equation $g(\xi, z) = 0$. We can assert the following.

Theorem 3.2. (i) *The variables x_j obeying system (18) preserve their ordering at all times, although some may converge to coincidence as $t \to \infty$.* (ii) *For the equilibrium to be stable it is necessary and sufficient that either* (a) *all α_j are strictly negative, or that* (b) *exactly one α_j is positive and*

$$\sum_j (\beta_j \gamma_j / \alpha_j) > 1. \tag{3.20}$$

The proof is given at the end of the section.

Consider the particular case

$$\dot{x}_j = -x_j + w_1 f(x_j) - w_2 \sum_k f(x_k), \qquad (3.21)$$

corresponding to $g(\xi, z) = -\xi + w_1 f(\xi) - w_2 z$. This can be put in the neural form (14) if f is sigmoidal. Suppose in fact that f is convex, so that for fixed z the roots x_j of (19) can take only the values a or b indicated in Fig. 3.3. By the second assertion of the theorem, at most one x_j can equal b, and, by assertion (i), this corresponds to the greatest y_j. Identification has thus been achieved in this case. It is also possible that all roots take the lower value a; this corresponds to the case when no element of y is larger than the others by a sufficient margin.

If f is sigmoidal then we are back in the case of Fig. 3.3 exactly, and the equilibrium x_j distribute themselves between the three roots a, b and c, with at most one equal to b. That is, the initial y-values are effectively categorised into two sets, the 'below threshold' and 'above threshold' values. Grossberg sees this as 'contrast enhancement', an important effect in visual processing.

Grossberg suggests a shunted version of (21), corresponding to $g(\xi, z) = -\xi + w_1(B - \xi) f(\xi) - w_2 \xi_z$. The same conclusions hold.

Another system would be

$$\dot{x}_j = -x_j + w \frac{\phi_j f(x_j)}{\sum_k \phi_k f(x_k)}, \qquad (3.22)$$

where the ϕ_j are positive constants. This is of the form (18) if the ϕ_j are identical in value, when we would again draw the same conclusions. However, this system has maximum-seeking properties, even if the ϕ_j differ, because, as we shall see in Chapter 8, it is derived from a steepest ascent principle. Note that the inhibitory/standardising effect is achieved by a division, rather than by subtraction plus shunting.

Proof of Theorem 3.2. (i) It follows from relations (18) that the rate of change of $\Delta = x_j - x_k$ is proportional to Δ if Δ is small. The difference may thus either grow or diminish exponentially fast in this region, but cannot change sign.

(ii) Let $\xi_j s^{st}$ denote a perturbation from the value of x_j at an equilibrium. It then follows from the linearised version of (18) that

$$s\xi_j = \alpha_j - \beta_j \sum_k \gamma_k \xi_k. \qquad (3.23)$$

We can eliminate ξ to deduce that s must satisfy the equation $H(s) = 0$, where

$$H(s) = 1 + \sum_k \frac{\beta_k \gamma_k}{s - \alpha_k}.$$

For stability we require that all roots of $H(s) = 0$ should have strictly negative real part. Graphing the function $H(s)$, remembering that $\beta_j \gamma_j > 0$, and assuming the α_j distinct, we see that it has real zeros between consecutive α_j and another below the smallest α_j, which accounts for all n roots. (If one of the α-values is p-multiple then there is a $(p - 1)$-fold s-root at that value.) We thus see that not more than one α_j can be positive, and that the greatest s-root can be negative in this case only if $H(0) < 0$ (since H has negative gradient between any pair of consecutive α-values). $\qquad\square$

Chapter 4

Bifurcations and Chaos

In treating a neural net one is considering a dynamic system. This is particularly the case for oscillatory nets. Dynamic systems theory has leapt ahead in recent years, and our discussion will barely meet material now regarded as standard. However, it is as well to clarify notation, some basic material and our general stance at this point.

4.1. The Hopf bifurcation

The canonical dynamic system in continuous time is specified by a set of n first-order differential equations in a vector variable x of dimension n:

$$\dot{x} = g(x, \gamma). \tag{4.1}$$

Here γ is a parameter of the system and $\dot{x}$ denotes the time differential dx/dt, which we shall sometimes also write as $\mathcal{D}x$. There is considerable interest in the equilibrium behaviour of such a system. The simplest equilibrium is a static one, at a point $\bar{x}$, say, which must then satisfy $g(\bar{x}, \gamma) = 0$. A given system may have many such equilibria.

The *domain of attraction* of the equilibrium $\bar{x}$ is the set of initial values $x(0)$ for which the solution $x(t)$ of (1) converges to $\bar{x}$ with increasing t. The equilibrium $\bar{x}$ is *locally stable* if this domain contains at least some n-dimensional neighbourhood of $\bar{x}$. There is considerable interest in the manner in which the pattern and nature of the equilibria change as the parameter γ is varied; catastrophe theory provides a codification of the types of change which are possible. Bifurcation theory tackles corresponding problems in cases where some of the equilibria may have a dynamic character, e.g. the periodic behaviour of a *limit cycle*, or the more complicated aperiodic or chaotic behaviours considered in the next section.

The time-honoured test of the character of a static equilibrium point is that of linearisation. If $\xi(t)$ is a small time-dependent perturbation of $x(t)$ from $\bar{x}$ then this will obey the linear equation

$$\dot{\xi} = G\xi \tag{4.2}$$

to within a term $o(\xi)$. Here G is the $n \times n$ matrix of first-order differentials $(\partial g_j(x)/\partial x_k)$, evaluated at $\bar{x}$. The general solution of (2) will be a linear combination of terms of the form ξe^{st}, where ξ is a constant vector and s a constant scalar satisfying

$$s\xi = G\xi.$$

That is, s is an eigenvalue of G and ξ is the corresponding right eigenvector. (We slur over the case of repeated eigenvalues, which can always be regarded as one of confluence.) Under mild regularity conditions, it is necessary and sufficient for local stability of $\bar{x}$ that all these eigenvalues should have strictly negative real part; i.e. should lie in the left half, $\text{Re}(s) < 0$, of the complex plane.

We should properly have written G as $G(\gamma)$, for the matrix will depend upon the parameter γ. Let us assume for simplicity that γ is scalar. It may be that for small enough γ all eigenvalues of $G(\gamma)$ lie in the left half-plane, but that, as γ increases, one pair of conjugate complex eigenvalues crosses the imaginary axis. That is, at a critical value γ_c of γ there is a pair of purely imaginary eigenvalues $s = \pm i\omega$. These give rise to a solution of (2) with the time-dependence $e^{\pm i\omega t}$; i.e. to a periodic solution of frequency ω.

Under some subsidiary conditions this burgeoning oscillatory instability marks the passage from a stable static equilibrium to a stable limit cycle; the transition is then termed a *Hopf bifurcation*. These are termed *transversality conditions*, meaning merely that they apply at a transition point. One condition is simply that further increase of γ should carry the root-pair into the right half-plane, in that $\text{Re}[ds(\gamma)/d\gamma] > 0$ at γ_c, where $s(\gamma)$ is one of the relevant eigenvalues. This will always be satisfied in the cases with which we shall be concerned.

The other condition is also a local one, designed to ensure that, despite apparent instability when $s(\gamma)$ develops a positive real part, nonlinearity of g limits the amplitude of the oscillation. Under such a condition the amplitude of the oscillation is typically of order $\sqrt{\gamma - \gamma_c}$, where γ_c is the critical value, for which $s(\gamma_c) = i\omega$. However, it must be said that this condition takes a truly horrific form when expressed in terms of the derivatives of $g(x, \gamma)$ (see e.g. Glendinning (1994) p. 227).

In the cases with which we are concerned the system (1) rather takes the form

$$\mathcal{L}x = g(x, \gamma), \tag{4.3}$$

where $\mathcal{L}$ is the constant-coefficient differential operator (i.e. a polynomial in $\mathcal{D}$) introduced in Section 3.4 and fully specified in Chapter 11. It has a stable inverse in that the solutions of $\mathcal{L}x = 0$ tend exponentially fast to zero with increasing t. In fact, $\mathcal{L}$ is only of order 2, so (3) could be reduced to the form (1), but there is no virtue in doing so.

Furthermore, in the cases we shall encounter, the function $g(x, \gamma)$ of (3) is bounded. Amplitude of any variation is then restricted, not by local conditions, but by the stability of the operator $\mathcal{L}$ plus the boundedness of g. In fact, in some cases (see Section 11.6), the transition at the bifurcation is continuous, in that the amplitude of the oscillation is indeed of order $\sqrt{\gamma - \gamma_c}$ for small positive $\gamma - \gamma_c$. In others (see Section 12.5) it is discontinuous, in that there is a range of γ for which both the static equilibrium and the limit cycle are stable, with the consequence that a limit cycle of substantial amplitude suddenly appears as γ increases through γ_c.

4.2. Chaos

One can view and define chaotic variation in several ways. Informally, it is continuing variation (usually in time) which is too complex that one would expect to be able to explain it by a simple model. It may seem to have a pattern (e.g. a tendency to periodicity) but the pattern is never perfect. One seems forced to explain the variation by the sustained injection of randomness, but that is no explanation. A typical electroencephalogram, such as that presented in Fig. III.1, would seem to represent just such a case.

However, there is nothing surprising if a complex system generates complex variation, and the neural system is certainly complex. In such cases one would say that the question had changed. Instead of asking 'How could the system have generated such a complex signal?' one should ask 'How can the system operate well when its signals are so complex?'. Nevertheless, the driving force behind the study of chaos over recent decades has been the realisation that simple models *can* generate variation which is of just such a character. That is, variation which continues indefinitely in a complex fashion, revealing a pattern, but one that is never perfectly repeated. One should have some feeling for the mechanism of such examples.

Note first that a finite-*state* machine can never produce chaos, since the output of an autonomous N-state machine will, after less than N steps, become periodic with a period not exceeding N. Chaos is then dependent on being able to distinguish infinitely many states. However, *finite-dimensional* dynamics can certainly produce chaos, in that a first-order recursion

$$x_{t+i} = g(x_t) \tag{4.4}$$

in terms of a continuous variable x can produce a chaotic sequence. The classic example, simple but archetypal, is the case when x is scalar and

$$g(x) = 2x \quad (\text{mod } 1) \tag{4.5}$$

whose graph is given in Fig. 4.1. Suppose that x_o has the binary expansion $0.c_1 c_2 c_3, \ldots$, so that

$$x_0 = \sum_{j=1}^{\infty} 2^{-j} c_j,$$

where the c_j are all either 0 or 1. Then one finds that, under the recursion (4), (5),

$$x_t = \sum_{j=1}^{\infty} 2^{-j} c_{t+j} = 0.c_{t+1} c_{t+2} c_{t+3} \cdots .$$

From this it follows that, if the sequence $\{c_j\}$ has period T, then so has the sequence $\{x_t\}$; if the sequence $\{c_j\}$ is aperiodic, then so is the sequence $\{x_t\}$.

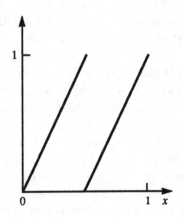

Fig. 4.1.   The stretch-and-truncate operator of equation (4.4).

In other words, the recursion (4) with choice (5) for g has 2^T solutions of period T (and possibly also shorter periods), for $T = 1, 2, 3, \ldots$, and infinitely many aperiodic solutions. We see whence this richness of variation springs: from the belief, possible only in a mathematical idealisation, that specification of x_0 implies specification of the whole sequence $\{c_j\}$. In a sense randomness *has* been injected, but in an infinitely convoluted and condensed form in the initial value x_0.

The operator (5) carries out the two operations of 'stretch' and 'fold', in that it first stretches x by a factor of 2 and then folds it by reducing it modulo 1. This seems to be the essential chaos-producing mechanism. The 'tent' operator depicted in Fig. 4.2, $g(x) = 1 - 2|x - \frac{1}{2}| (0 \leq x \leq 1)$, also has this property and also yields a chaos-producing recursion (4). The low-order systems of differential equations in continuous time with chaotic solutions also demonstrate the stretch-and-fold property. Of these the Lorenz equation, an early meteorological model, provides a classic example. (See e.g. Thompson and Stewart (1986), Glendinning (1994).)

Note a consequence of the stretch operation: that solutions $x(t)$ which begin from neighbouring values of $x(0)$ will diverge from each other at a rate exponential in t. Note also that, if one in fact allows the dynamics to be driven by noise, then it is the combination of the operations of 'stretch' and 'noise injection' which produces power-law autocorrelation and inverse power-law spectral densities (Whittle, 1962).

While 'stretch and fold' is regarded as the archetypal chaos-generating mechanism by pure mathematicians, physicists are moved by other modes of thought. Consider a model of a unit mass with coordinate x in $\mathbb{R}^n$,

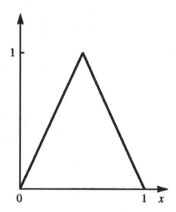

Fig. 4.2.   The tent function — a stretch-and-fold operator.

moving under Newtonian dynamics in a force-field derived from a potential function $V(x)$. Then x obeys the equation

$$\ddot{x} = -\frac{\partial V}{\partial x} \tag{4.6}$$

where $\ddot{x} = \mathcal{D}^2 x$. Suppose the mass has been captured in a parabolic potential well situated at the origin, so that $V(x) = \frac{1}{2}x^T W x$ for some positive definite matrix W. Equation (6) then becomes

$$\ddot{x} + Wx = 0 \tag{4.7}$$

with solution

$$x(t) = \sum_j c_j \xi_j \sin(w_j t - \phi_j). \tag{4.8}$$

Here ω_j^2 is an eigenvalue of W and ξ_j is the corresponding normalised eigenvector; the amplitudes c_j and phases ϕ_j are determined by initial conditions. The motion is thus a sum of sinusoidal components of frequencies ω_j, corresponding to the 'modes' of the linear system (7). If the frequencies are not mutually commensurable (i.e. integral multiples of some basic frequency) then $x(t)$ is aperiodic. However, one will still not regard it as chaotic, since it has the finite representation (8). This would be revealed physically by its power spectrum: concentrated on the frequencies ω_j. Furthermore, the solution depends continuously (uniformly in t) on the initial values.

However, suppose that $V(x)$ departs from the paraboloidal form as $|x|$ increases, so that to equation (7) should be added nonlinear terms of order $o(x)$ for small x. These will cause the modes to interact so that, crudely speaking, all frequencies of the form $\sum_j r_j \omega_j$ will occur, where the r_j are integers of either sign. If the ω_j are incommensurable then all frequencies can occur (with emphasis on the naturally-occurring ω_j), so that the motion will appear chaotic.

Actually, the final outcome is not easy to quantify. With transfer of energy between modes now possible, energy may drain away from some modes altogether. If only one mode were left, then the motion would be periodic.

Such an attrition of modes could mean that $(x, \dot{x})$ ends up in a set $\mathcal{A}$ of smaller dimension (in some curvilinear sense) than the original $2n$: an *attractor*. However, the degrees of freedom still present in the solution plus the curvilinearity of $\mathcal{A}$ may still leave the pattern of variation a non-recurrent one.

A Newtonian model such as this has the special feature that the amplitude of motion is determined by the initial energy, which is conserved. In other words, motion is confined to a shell of constant energy, of which there are a continuous infinity. In very many applications of interest a model will show only finitely many attractors, and the attractor dictates the amplitude of motion.

Part II

Associative and Storage Memories

Chapter 5

What is a Memory? The Hamming and Hopfield Nets

5.1. Associative memories

In everyday contexts a 'memory' can mean either the mechanism that does the remembering or the thing remembered. In technical contexts the first meaning is generally the one understood; the term 'memory device' can be used if one wishes to be more explicit. The thing remembered is referred to as a 'memory trace'.

One particular memory device is the *associative* or *content-addressable* memory. This has a number of memory traces built into its structure, and is so designed that, on being presented with data from the external world, it will select that trace which seems to match the data best in some sense. The classic situation is that of character recognition. Suppose that the traces are just the 26 characters of the standard English alphabet, and that the image presented (the data for a particular trial) is a version of one of these characters. It is a 'version' in that it may be subject to all the variations of size, font and orientation and all the distortions introduced by the individual hand of a writer or the 'noise' of faulty data transmission. The associative memory embodies a recognition rule, which decides which character was intended. There will be errors, of course, if distortion is such that it can create real ambiguity.

This is a simple and formalised version of the much more general problem: that of 'making sense' of sensory input from the external world.

Suppose there are p traces; let us label them a_j ($j = 1, 2, \ldots, p$). For the character recognition example these would be standard versions of the $p = 26$ characters of the alphabet. More abstractly, we could label them by their list-places j; the character 'd' thus corresponds to $j = 4$. Let us denote the input data by y; for the recognition problem this would be the

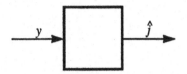

Fig. 5.1.   A block representation of an associative memory.

distorted image which is presented. Then the associative memory could be regarded as a device with input y and output $\hat{j}$, where $\hat{j}$ is an *estimate* of the label of the true trace. We can represent the device by the block diagram of Fig. 5.1; such diagrams acquire more point as we move on to details of structure.

A particular associative memory is then specified by the rule which determines output $\hat{j}$ in terms of input y, and we should like to choose this rule so that the proportion of cases which are misclassified is as small as possible, in some sense. A statistician would of course recognise the problem as one of *hypothesis testing*. There are p hypotheses, exhaustive in that one of them must be true and mutually exclusive in that only one can be true. One wishes to infer which is true in a particular case on the basis of data y. Statistical theory derives inference rules which are optimal under specific statistical assumptions on the *model*, i.e. on the nature of the 'signals' a_j and the distortion which they suffer in production of the image y. One will in general not be able or willing to make such assumptions in the neural context, but statistical theory at least provides the best guide for inference rules which are likely to perform well.

One model which may be very far from reality in many cases but which provides a firm starting basis for the development of ideas is the *linear model*. Suppose that the traces a_j and the data image y are all column n-vectors, obeying the relation

$$y = a_j + \epsilon \qquad (5.1)$$

if y is indeed a corrupted version of the jth trace. Here ϵ is the statistical 'noise' which distorts the trace, in this case by simple addition. One can regard y as the data derived from an array of n point sensors ('pixcells'), and a_j as the data which would have been received if the jth trace had been presented without distortion.

We can more conveniently write model (1) as

$$y = Ar + \epsilon \qquad (5.2)$$

where A is the $n \times p$ matrix whose columns are the traces $a_k (k = 1, 2, \ldots, p)$ and r is a column p-vector whose elements are all zero except for a unit in the place corresponding to the trace actually present — place j in case (1). Estimation of j is then equivalent to estimation of the *coefficient vector* r under this constraint.

An *autoassociative memory* gives as output the actual trace $a_{\hat{j}}$ for which it has decided rather than simply the label $\hat{j}$ of that trace. So, if we regard y as a sensory input, then $a_{\hat{j}}$ is of the same character, in that it is a sensory image of the inferred trace rather than just an abstract label for it. In the linear case we can write this output as $A\hat{r}$.

5.2. The Hamming net

The problem of determining which trace a_j gave rise to an observation vector y through the relation $y = a_j + \epsilon$ is one which has been developed so much in the signal and communication context that it is worthwhile to make the connection. In this context a_j would be regarded as the 'transmitted signal' (a signal over a block of n consecutive instants of time) and y as the 'received signal', a version of the transmitted signal which has become distorted during passage through the communication 'channel'. The p possible forms of the transmitted signal (the 'codewords' a_j) are known at the receiver, and it is a question of inferring from y which one of these was in fact sent.

Suppose that the memory traces have been standardised to a common amplitude $|a_j|$ (where $|a_j|^2 = \sum_k a_{jk}^2$). Then, under a variety of assumptions (see the notes below), the maximum likelihood solution of the problem is to take the estimate $\hat{j}$ as the value of j maximising the inner product $a_j^T y$. (In the communication context this inner product is generated as the output of a *matched filter*.) We could write this as

$$\hat{r} = M(A^T y) \tag{5.3}$$

where M is the operation which replaces the maximal element of a vector by unity and all others by zero. This inference rule seems to be have been proposed in the coding context first by Hamming (see the references in Gallager (1968), Lippmann (1987) and Baum, Moody and Wilczek (1987)); it was certainly proposed early by Steinbuch (1961), Taylor (1964) and Grossberg (1976a). We shall then refer to the net which realises the rule as the *Hamming net*. Despite being derived by statistical optimisation it has a neural look about it: a linear operation A^T followed by the maximisation operation which we know can be realised neurally (see Section 3.6).

A simplification carrying some slight risk would be to replace the maximising operation in (3) by a threshold operation, which replaces each element in the vector $A^T y$ by a unit or zero according as that element does or does not exceed a prescribed threshold value x^*. Under appropriate statistical assumptions the probability is high for large n that the threshold will be exceeded for exactly one value of j, the correct one, if x^* is appropriately determined. One might write this modified inference rule as

$$\hat{r} = f(A^T y) \qquad (5.4)$$

where f is a hard threshold function with threshold x^*, and we understand from (4) that this operation is to be applied to each element of the vector separately. Relation (4) certainly has a neural look to it. We shall refer to it as the *threshold Hamming net*. (See Whittle (1989) and Meilijson, Ruppin and Sipper (1995).)

The Hamming rule (3) tends to be passed over in the ANN literature, partly because it is so simple that there is seemingly not much that one can say about it, partly because it is so obviously derived from a very special model, partly because it is not so easily seen as a rule which can be 'learned'. However, it is in fact integral to the 'feature mappings' of Amari and Kohonen (see Chapter 10) and its inevitability is emphasised by its unforced appearance, in dynamic form, in the solution of the fading data problem (see Chapter 8).

Notes

1. The *memoryless Gaussian channel*. Suppose that (1) holds and that the elements of the noise vector are independently and normally, distributed with zero mean and variance v, so that the probability density $\psi_j(y)$ of y for a prescribed value of j is proportional to $\exp[-|y - a_j|^2/2v]$. Suppose also that the trace a_j has a prior probability of π_j. Then the minimum expected error rule is to choose the value of j maximising $\pi_j \psi_j(y)$ (see note 2 of Section 2.1). This is the value of j maximising $a_j^T y + v \log \phi_j$, where $\phi_j = \pi_j \exp[-|a_j|^2/(2v)]$. The rule reduces to the Hamming rule (3) in the case when the traces are equiprobable and standardised.

   In the communication context one would indeed refer to the 'channel' introducing the distortion as 'Gaussian' because of the normality assumption, 'memoryless' because of the independence assumption.

2. The *memoryless binary channel*. Suppose that the elements of both the memory traces a_j and the data vector y take the values ± 1. Suppose that, if the actual trace present is a_j, then y_k differs from a_{jk} with probability μ ($k = 1, 2, \ldots, n$), these errors being statistically independent for differing k. Then the probability that the data vector takes the value y is $\mu^{d_j}(1 - \mu)^{n-d_j}$, where d_j is the number of errors, i.e. the number of places in which a_j and y differ. If $\mu < \frac{1}{2}$ then the maximum likelihood rule (or the minimum error rule if the traces are equiprobable) will infer the value of j which minimises d_j. But $a_j^T y = n - 2d_j$, so this is just the Hamming rule. The reason for the description 'memoryless binary channel' will be clear.

5.3. Autoassociation, feedback and storage

The autoassociative version of the threshold Hamming net (4) has output

$$\hat{y} = AfA^T y. \tag{5.5}$$

Here we have dropped brackets, and are regarding f simply as a nonlinear operator with the effect that, if x is a vector with elements x_j, then fx is a vector with elements $f(x_j)$. We have also denoted the output by $\hat{y}$ to indicate that it is a 'categorised' version of the input y — categorised, in that y has been reduced to one of the stereotypes a_j.

Consider a general case, for which the data array and the traces a_j take values in a space $\mathcal{Y}$ and the autoassociative memory has action

$$\hat{y} = gy. \tag{5.6}$$

Here g is an operator, the *autoassociative operator*, in general nonlinear. The minimal property one would demand of such a memory would be that, if fed with a pure trace, it would yield that same trace as output. In other words, the traces are all fixed points of the operator g:

$$ga_j = a_j \quad (j = 1, 2, \ldots, p). \tag{5.7}$$

In fact, one would demand rather more: that the memory would yield the output a_j for *any* input in a reasonable neighbourhood $\mathcal{N}_j$ of a_j:

$$gy = a_j \quad (y \in \mathcal{N}_j; \ j = l, 2, \ldots, p). \tag{5.8}$$

A weaker version of this demand is to require that *iteration* of g would bring any element of $\mathcal{N}_j$ to a_j ultimately. That is, we require that the traces a_j

should be equilibrium points under the recursion

$$y_{t+1} = gy_t, \quad (t = 0, 1, 2, \ldots), \tag{5.9}$$

and that $\mathcal{N}_j$ should lie in the domain of attraction of a_j $(j = 1, 2, \ldots, p)$. Specifically,

$$y_t \to a_j \text{ as } t \to \infty \text{ for any } y_0 \text{ in } \mathcal{N}_j \quad (j = 1, 2, \ldots, p). \tag{5.10}$$

The Hamming operator AMA^T of course has the stronger property (8): of attracting an input vector to a pure memory trace in a single step. However, there are good reasons, already mentioned, for considering memories for which this attraction takes place progressively over time. One is that of achieving an extremal operation by natural dynamics; the other that of facing the 'fading data' problem of Chapter 8. One might describe y_t as an 'idealised' data array, which represents a progressive transformation from the raw value $y_0 = y$ to a terminal value identical with a pure trace.

By modifying the one-step relation (6) to the recursion (9) we are in effect introducing feedback from output to input. This has the effect of producing a device whose state variable y_t, once in $\mathcal{N}_j$, stays in $\mathcal{N}_j$. That is, it is a *storage memory* in that, once set in a given 'memory state' corresponding to the subset $\mathcal{N}_j$ of physical states, it stays there. This may seem a fairly trivial notion. After all, why not simply postulate the relation $y_{t+1} = y_t$, and achieve as many memory states as there are physical states? In fact, there are two reasons, both concerned with 'noise' (i.e. stochastic error), and they imply that a memory state can only be associated with a substantial basin of attraction.

The first is that of *observation noise*, i.e. of sampling error in the original y. The neighbourhoods $\mathcal{N}_j$ of (10) must be large enough (i.e. the basins must be wide enough) that y is captured in the correct basin with high probability. The second reason is that of *process noise*, i.e. of stochastic disturbance to the recursion (9) itself. The attraction to the relevant equilibrium must be strong enough (i.e. the basins must be deep enough) that y, once captured, is retained with high probability for a sufficient time.

To summarise, then: any autoassociative memory can be converted to a storage memory by the introduction of feedback. It is the intrinsic property of the autoassociative memory (that all the memory traces are local points of attraction) which gives the storage memory corresponding basins of attraction and reliability of storage. The question is then one of quantitative

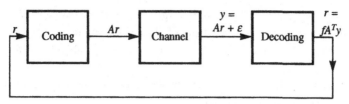

Fig. 5.2. A block diagram of the closed communication link (the antiphon) which realises the autoassociative Hamming net. Note that r is an estimate of the loading vector rather than the true value; its value will stabilise after one iteration if there is no noise in the loop. The value of y will vary with iteration if there is noise in the loop; the value of r should nevertheless be stable over many iterations. Simulations show it to be extremely stable.

performance; for given physical resources and reliability demands, how many memory states (i.e. basins of attraction) can be achieved?

The storage memory derived in this way from the threshold Hamming rule (5) is just the *antiphon* proposed by the author (Whittle, 1989, 1990, 1991a) to demonstrate reliable memory (i.e. memory storage) in the presence of noise. The picture envisaged was exactly that of transmission through a noisy communication channel with feedback, in that the decoded output is resubmitted as input: see Fig. 5.2. Specification of the message to be transmitted amounts to specification of r; if it takes the value e_j then the jth message value is intended. (Here e_j is the p-vector with a unit in the jth place, zeros elsewhere.) In the coding step r is transformed to Ar, and so e_j to a_j, which is then the 'signal' transmitted. This is distorted in the noisy channel to $y = Ar + \epsilon$. In the decoding step the true value of r is estimated from channel output y as $\hat{r} = fA^Ty$, by the threshold Hamming rule. The whole system is then made into a storage memory by feedback, i.e. by taking output $\hat{r}$ as the new message value r to be encoded and transmitted. See also Mazza (1997).

The autoassociative operator AfA^T characterises what we may term either the threshold Hamming rule or the antiphon rule. Because we see continued application of the rule as a feedback, this operator is equivalent to cyclically permuted versions of it; to A^TAf or fA^TA. These just correspond to changes of variable in the recursion (e.g. from y to A^Ty), or to the reckoning from different starting points in the feedback loop.

The values $r = e_j$ (corresponding to values a_j for $y = Ar$) are indeed stable equilibrium points of the noiseless system. Both analysis and computational experiment reveal them to be long-lived as equilibria even in the presence of noise. But, of course, the question is again one of quantitative performance. How large can p (the number of memory traces) be for a given specification of dimension n, noise characteristics and reliability demands? This is a question we consider in Chapters 6 and 9.

The recursion $y_{t+1} = AfA^T y_t$ for the noiseless system can be decomposed into the pair of recursions

$$r_{t+1} = fA^T y_t \tag{5.11}$$

$$y_{t+1} = Ar_{t+1}. \tag{5.12}$$

These can be viewed as an alternation between an 'abstract' and a 'sensory' version of the message, represented by r and $y = Ar$ respectively. The implicit presence of such an alternation is very common in ANNs — an alternation between a reduced representation and a reconstruction $\hat{y}$ deduced from it. We see it in the principal component techniques, the Helmholtz machine and Grossberg's adaptive resonance devices described in Chapter 10. It is the reason for choice of the term 'antiphon': a piece in which verse and response are alternately sung by each of two choirs.

All these devices demonstrate the feature which von Neumann (1956) saw to be necessary for reliable operation in the presence of noise: what he termed a 'restoring organ'. This means in fact stabilisation of a wandering state variable by attraction to the nearest of a number of centres. In the present case this restoration is provided by the inference step, which attracts a state variable starting from $y_0 = y$ to the nearest a_j, either in a single step or progressively over time.

5.4. The Hopfield net

The Hopfield net was one of the first associative memories of a distinctly neural character. Its proposal by Hopfield (1982) opened a new epoch, established the 'Hopfield net' as one of the central ANN concepts, and stimulated an ever-growing literature. In fact, the net already had convincing antecedents, in that versions of it had been proposed by Nakano (1972), Kohonen (1972) and Anderson (1972) and analysed by Amari (1972, 1977a). However, as is common, Hopfield's (independent) proposal came when the time was right, so that it rode on the crest of the wave that it generated. It was also given special impetus by the realisation of its connection with the models and methods of statistical physics: see e.g. Amit *et al.* (1985), Amit (1989), Gardner (1986) and Kamp and Hasler (1990).

A general set of n linear mutually-connected linear gates in discrete time would obey the equation

$$y_{t+1} = f(Wy_t), \quad (t = 0, 1, 2, \ldots) \qquad (5.13)$$

Here y_t is the vector of outputs of the n nodes of the net at time t, and W is the synaptic matrix, whose jk^{th} element w_{jk} is the coefficient of the output of node k in the input to node j.

For the Hopfield net the data vector is taken as the initial value, $y_0 = y$, and the synaptic matrix is related to the matrix of traces by $W = AA^T$ (with some variants under different proposals). The activation function f is usually taken as the hard threshold function $f(\xi) = \text{sgn}(\xi)$, in which case W needs no scaling and the elements of y_t take the values ± 1. If a more general activation function is used then W is usually scaled by a factor p^{-1}, either explicitly or by incorporating this factor in the definition of f.

We see similarities with and differences from the Hamming net. For the Hamming net the autoassociative operator g is AfA^T; for the Hopfield net it is fAA^T. Neither can be reduced to the other by the cyclic permutations induced by a change of variable. They can be brought respectively to the forms $A^T Af$ and $AA^T f$, so one might say that the Hamming net acts as a Hopfield net for the reduced description rather than for the full description. We note that the Hamming net has a clear derivation, and wonder what that of the Hopfield net might be. Another difference is that the Hamming net gives its final output after a single step (at least if we take the strict selection rule of $g = AMA^T$), whereas the Hopfield net in general takes several steps before it even approaches a pure trace.

One sees how the net operates. Consider its first step, when the operator fAA^T is applied to y. The formation of $A^T y$ achieves the same dimensional reduction as does the Hamming net: to a set of p linear forms. The basic consistency demand (7) implies that $AA^T a_j$ must be approximately a scalar multiple of a_j for each j; i.e. that the traces are approximately eigenvectors of $W = AA^T$. This can be true only if the traces are approximately mutually orthogonal — a very restrictive condition.

There is another interesting point. In the next chapter we shall consider the question of performance, and shall see that, for large n, an associative memory should be able to deal reliably with exponentially many (in n) memory traces. However, both numerical experiment and mathematical analysis reveal the necessity of a linear bound on p if operation of the Hopfield net is to be reliable. In fact, the ratio $\alpha = p/n$ should not exceed a critical value of about 0.138 in the limit of large n (less if f is soft or if there is process noise in the system).

We shall find one reason for this apparent contradiction in the next chapter, and shall then review the Hopfield net more thoroughly. A more fundamental explanation will emerge in Chapter 8.

However, the reader may be wondering already why the net should have exerted the fascination that it has done, considering that it shows neither the simple rationale nor the simple operation of the Hamming net, and that it is actually useless for all but almost-orthogonal traces. The net certainly has the properties that it is represented by a McCulloch-Pitts net as it stands, and that it can be 'learned' (see Chapters 6 and 10). However, the fascination it induces is perhaps because it also incorporates another function, one which is vaguely felt rather than explicitly identified. We identify that function in Chapter 8.

5.5. Alternative formulations of the Hopfield net

If the threshold function f of (13) is invertible then one can use the relation $y = f(x)$ to define a transformed variable x in terms of which equation (13) becomes

$$x_{t+1} = Wf(x_t). \tag{5.14}$$

A continuous-time version of this equation would be something like

$$\dot{x} = Wf(x) - x. \tag{5.15}$$

Relation (15) contains a nonlinear positive feedback term, intended to guide the variable towards an equilibrium corresponding to the nearest pure trace, and a linear negative feedback term, which assures stability of these equilibria.

If we replace the linear gate by the biologically more faithful version of a neuron described in Section 3.4, then (15) would be replaced by

$$\mathcal{L}x = Wf(x). \tag{5.16}$$

Stability at equilibria is now encouraged by the intrinsic stability of the linear operator $\mathcal{L}$, and assured if appropriate conditions are placed on W (see Chapter 12). In this context we would regard x as the vector of membrane potentials of the neurons and $y = f(x)$ as the vector of consequent firing rates. Relation (16) then describes an assembly of neurons with a mutual feedback specified by the synaptic matrix W.

It is convenient to term any McCulloch-Pitts net with symmetric connections an 'H-net'. This then embraces both the Hopfield net and the autoassociative Hamming net (to within a variable transformation). These two behave very differently, as we know, and are not in the least to be confused. However, it is useful to have a neutral term which covers both.

Chapter 6

Compound and 'Spurious' Traces

6.1. Performance and trace structure

The *scale* of a memory device can conveniently be measured by n, the dimension of the data vector, and its *size* by m, the number of memory traces which the device can reliably distinguish (in the case of an associative memory) or hold (in the case of a storage memory). We denote the total number of traces by m rather than p for a reason which soon transpires. Reliability assertions generally take the form of limit statements for large n.

The performance question is: how quickly can m grow with n, consistent with reliability, for given physical resources and for a given statistical characterisation of traces and noise? Statistical communication theory suggests that, under reasonable assumptions, m should be able to grow exponentially fast with n. However, it turns out that such performance can be realised in general only if the amount of computation (and so the physical plant devoted to implementing it) also grows exponentially fast with n. This is unacceptable. On any reasonable understanding of performance one would expect plant to scale no faster than some power of n; preferably, to scale linearly.

It turns out that exponential growth of size and reasonable growth of plant can be reconciled if the m trace vectors have some structure. A particular such structure, perhaps representative of all, is that in which one derives $m = 2^p$ traces by forming the 2^p linear combinations (with coefficients 0 or 1) of p *basis traces*.

This is a reasonable structure in the biological context, and indeed that which one would expect. In the olfactory context one set of odours can very well simply be added to another. In the visual context the superposition is not linear: 'cat on the mat' is not quite the same as 'cat plus mat'.

Nevertheless, the notion that simple images are compounded in some rather direct fashion is very evident.

The alternatives of simple versus compound traces correspond to what one might term the 'grandmother' and 'factorial' formulations. A 'grandmother' cell is one tuned to recognise one fully-specified image (the grandmother, in the classic example) which is not decomposed into constituent features. This is characteristic of the simple trace. If one admits decomposition into features which one might regard as more elementary ('factors'), then one accepts the idea of a compound trace.

By following up these ideas we shall develop the implications of the simple linear model (4) below and elucidate some of the puzzling features of the Hopfield net.

6.2. The recognition of simple traces

Consider the recognition of one of m traces by the threshold Hamming rule (5.4), which we know to be close to the maximum likelihood rule. The traces are simple, in that just one of the elements r_j of the loading vector r should be unity, all the others zero.

One can relatively easily derive an upper bound for $P_j(\text{error})$: the probability that an error is made in estimation, given that the true trace was a_j. Consider, for example, the simple linear model $y = a_j + \epsilon$, for which y has probability density $\psi_j(y) \propto \exp(-\frac{1}{2}|y - a_j|^2/v)$ under the Gaussian assumptions of Section 5.2. One finds the bound (see the note)

$$P_j(\psi_k > \psi_j) \leq \exp\left(-\frac{1}{8}|a_j - a_k|^2/v\right)$$

for the probability that $\psi_k(y)$ is greater than $\psi_j(y)$. This then implies the bound

$$P_j(\text{error}) \leq \sum_{k \neq j} \exp\left(-\frac{1}{8}|a_j - a_k|^2/v\right) \tag{6.1}$$

if the m traces are equiprobable (when the maximum likelihood rule is optimal).

However, expression (1) is not easily handled as it stands. The symmetrising procedure of random coding which is so effective in statistical communication theory now has the analogue: suppose that the traces a_j are chosen randomly and independently from some set $\mathcal{A}$, and consider error probabilities under random variation of traces as well as of noise. In the present case, suppose that the elements a_{jk} are independently and normally distributed with variance P. The prescription of variance implies a power constraint on the signal: that $E|a_j|^2 = nP$. Averaging expression (1) over trace variation we have then

$$\overline{P}_j(\text{error}) < m \left[1 + \frac{P}{2v} \right]^{n/2}, \tag{6.2}$$

where the overbar indicates that the error probability is averaged over random choice of the traces. The bound (2) is independent of j, and so represents a bound on average probability of error, $\overline{P}(\text{error})$, quite simply.

By the same style of argument one can arrive at a bound of the type

$$\overline{P}(\text{error}) \leq me^{-Kn} \tag{6.3}$$

for a general class of cases, where K is some constant which depends upon the statistics of the particular case.

Reliability under some trace choices (in fact, under almost all choices for the trace statistics assumed) will be ensured if the averaged error probability (3) converges to zero with increasing n. This reliability condition is satisfied if m grows with n in such a way that $n^{-1} \log m$ has a limit value less than K for large n. One can then regard K as a lower bound on the inferential *capacity* of the system, in nats per element of y.

However, this rule requires m linear gates when realised by a network (see Section 5.2), and so at least m nodes. That is, the number of nodes in the network will also increase exponentially with n if we allow m to do so. As emphasised in the previous section, growth of scale at this rate is unacceptable. The problem is resolved by the imposition of some structure on the traces, so that the m vectors a_j are not chosen completely independently.

The reader may have doubts about the notion of random traces. In the communication context one can choose the poding of message to signal as one wishes, subject to physical feasibility, but the coding of 'entity' to 'sensory image' is not in general at the disposition of the network designer. However, one can say that the approach yields an evaluation of expected performance over a class of cases, which may be more meaningful than an evaluation of performance in a given case.

Note that the exponent $K = \frac{1}{2} \log[1 + \frac{1}{2}P/v]$ implied by (2) is not the best possible; see Section 9.3.

Note

In deriving the first inequality above we have appealed to a standard bound: that if x is a standard normal variable then $P(x \geq u) \leq \exp(-\frac{1}{2}u^2)$ for $u \geq 0$. This is a Chernoff bound, following from the argument

$$P(x \geq u) \leq \inf_{\theta \geq 0} Ee^{\theta(x-u)} = \inf_{\theta \geq 0} e^{-\theta u + \theta^2/2} = e^{-u^2/2}.$$

6.3. Inference for compound traces

Consider then a model which allows the basic traces a_j to be compounded linearly. This would be a realistic model in the olfaction context, when it is certainly natural to allow that several stimuli (odours from different sources) may be present simultaneously, in varying intensities. In the familiar linear model

$$y = \sum_j r_j a_j + \epsilon = Ar + \epsilon. \tag{6.4}$$

the loading vector r is then to be seen as the vector of intensities of the p possible basic stimuli. It can take any non-negative value, and the effective trace Ar is then genuinely compound. The value of r will vary from trial to trial (i.e. with observation at different times) and is in fact the state variable for the problem, to be estimated.

Later we shall consider the quantised case, when the elements of r can take only the values 0 or 1, so that the basic stimuli are either absent or present (at a standard intensity). For the moment we shall abandon constraint and allow r to take any value in $\mathbb{R}^p$. From the statistical point of view we are then considering the most basic and well-worn topic of the subject, linear regression, although with the twist that the parameter r varies randomly between trials. The idea of increasing information by repeated sampling then has to be replaced by that of acquiring more information at the time. That is, by placing more sensors, and so increasing n.

The solution under standard assumptions is well-known.

Theorem 6.1. *Suppose that the noise vector is normally distributed with zero mean and (for simplicity) covariance matrix $V = v_\epsilon I$. Then the maximum likelihood estimate of r is*

$$\hat{r} = (A^T A)^{-1} A^T y \tag{6.5}$$

and

$$E(\hat{r}) = r, \quad E[(r - \hat{r})(r - \hat{r})^T] = v_\epsilon (A^T A)^{-1}. \tag{6.6}$$

Proof. The first assertion follows by maximisation of the expression $\exp[-\frac{1}{2}v_\epsilon^{-1}|y - Ar|^2]$, proportional to the likelihood, with respect to r. The second follows from appeal to the relation $\hat{r} - r = (A^T A)^{-1} A^T A \epsilon$. $\quad\square$

Of course, no novelty is claimed for Theorem 6.1. To a statistician this is plain old regression, with a history going back to Gauss. In the ANN context it has been derived early (in ANN terms) and often, as an optimal 'linear associative memory'.

We see from (6) that the 'smaller' the matrix $(A^T A)^{-1}$ the smaller the estimation error (in mean square) and the hope would indeed be that this error would tend to zero with increasing n. A necessary condition is certainly that $A^T A$ should be nonsingular, which in turn implies that $p \leq n$. The ratio $\alpha = p/n$ will later turn out to be significant; we see the necessity for the condition $\alpha \leq 1$.

The autoassociative operator corresponding to the estimate (5) is

$$g = A(A^T A)^{-1} A^T \tag{6.7}$$

which has the effect of projecting y on the manifold $A\mathbb{R}^p$ generated by the p columns of A. It is idempotent (i.e. $g^2 = g$), as a projection operator should be — just a reflection of the fact that rule (5) achieves the desired estimate in a single step. Expression (7) is a matrix, but not a stability matrix, and its iterates would wander widely in the presence of computational noise. It would then not serve as an effective rule for a storage memory — restoration to the manifold $A\mathbb{R}^p$ is not a sufficiently powerful restoration. For stability we shall have to consider a quantised r and correspondingly quantised estimates, The most radical such quantisation which allows the stimuli to occur independently is to assume that the elements of r are restricted to the values 0 or 1, i.e. to restrict r to the set $\{0, 1\}^p$.

One could consider a maximum likelihood estimate restricted to such a quantisation, but the simplest course is to adapt the unrestricted estimate $\hat{r}$ to

$$\tilde{r} = f(\hat{r}) \tag{6.8}$$

where $f(\hat{r}_j)$ takes the values 0 or 1 according as $\hat{r}_j \leq \frac{1}{2}$ or $\hat{r}_j > \frac{1}{2}$.

Relations (5) and (8) represent again an associative memory, but one which we can expect will show stability. It is simple enough in concept; the term 'quantised regression' which we shall adopt for it describes it exactly. We shall study its performance in Sections 5 and 6. Note now that its implied autoassociative operator is

$$g = Af(AA^T)^{-1}A^T, \tag{6.9}$$

to be compared with the Hamming and Hopfield operators AfA^T and fAA^T. Note that reduction of dimension (from n to p) again occurs in the formation of $A^T y$.

6.4. Network realisation of the quantised regression

The quantised regression, expressed by relations (5) and (8), constitutes an associative memory designed to deal with compound traces, with the autoassociative operator (9). It has the virtue of being a one-step operator, not apparently needing recursion.

However, there is a submerged computational burden. Operator (9) indeed involves only linear and threshold operations, but it also involves a considerable calculation: the inversion of the matrix AA^T. This may well have to be tackled recursively.

The preliminary estimate $\hat{r}$ is determined by the linear equation system

$$(A^T A)\hat{r} = A^T y. \tag{6.10}$$

There are several iterative methods for solving such systems, but the natural one in the present context is the Gauss-Seidel method. Let us write $AA^T = W_0 + W_1$, where W_0 and W_1 are respectively the diagonal and non-diagonal parts of the matrix. Then we could replace equation (10) by the recursion

$$r_{t+1} = W_0^{-1}[A^T y - W_1 r_t], \tag{6.11}$$

where r_t is the solution for $\hat{r}$ yielded at the tth pass. The recursion will converge if $W_0^{-1}W_1$ is a stability matrix (i.e. has all its eigenvalues strictly inside the unit circle).

Note that recursion (11) differs in one significant respect from the autoassociative recursions we have considered. The data y enter, not in the initial value for the recursion, but in a term of fixed form which drives the recursion at all stages. This is attractive, in that it implies that the data are injected, not by a radical resetting of the state variable of the system, but by the gentler mechanism of a fixed input which gradually swings the state variable its way. On the other hand, there is an implied assumption: that the data can be stored without degradation — i.e. can themselves be 'remembered'. As we shall see in Chapter 8, this a crucial and questionable assumption.

6.5. Reliability constraints for the quantised regression

Performance of the unquantised estimate $\hat{r}$, given in (5), is expressed by relations (6), which say about as much as one can without further assumptions on A. We are now interested in performance of the quantised estimate $\tilde{r}$, given in (8), and in how rapidly p may grow with n, consistent with reliability.

The condition $p \leq n$, necessary for nonsingularity of $A^T A$, implies that linear growth of p with n is the best we can expect. The ratio $\alpha = p/n$ turns out to be significant. If reliability or other considerations forbade α to exceed some critical value α_c then this would still be consistent with the attainment of a positive capacity. The number of compound traces which could be distinguished would be

$$m = 2^{\alpha_c n + o(n)}$$

so that a capacity of least α_c bits per symbol could be attained. Furthermore, the number of nodes required by the network realising the quantised regression would then be $\mathcal{O}(n)$.

We have given the various interpretations of the problem, as one of distinguishing between compound versions of olfactory stimuli, memory traces or codewords. There is yet another: in forming the compound stimuli Ar for varying binary vectors r we are in fact forming the words of a linear code with basis matrix A. The inference problem is one of decoding the word after transmission through a noisy channel. There are p words a_j in the basis, so the ratio $\alpha = p/N$ can be appropriately termed the *basis rate*.

We shall evaluate performance by calculating a bound on the probability of error averaged over statistical variation of A. That is, we are effectively considering a linear code with a random basis.

There are at least three reasons for allowing randomness in the coding. One is that of mathematical expediency: averaging may reduce an unamenable error bound, such as (1), to an amenable one, such as (2). Another is that it is reasonable to assess performance over a class of cases rather than for a single case. Yet another consideration is the general philosophy behind artificial neural networks: that their construction is not by the exact realisation of an exact plan, but by a 'guided random' evolution. The view that such a looser approach can be effective is given support by the fact that the asymptotically optimal codes of statistical communication theory are realised randomly.

We shall now analyse explictly the case of Gaussian a and ϵ, for which certain relations hold exactly and permit a simple treatment.

Theorem 6.2. *Suppose that the errors ϵ_j are normally distributed with zero mean and variance v and that the letters a_{jk} are normally distributed with zero mean and variance P. Then the probability of error (averaged over codings) has the upper bound*

$$\overline{P}(\tilde{r} \neq r) \leq \alpha n[1 + P/(4v)]^{(1-\alpha)n/2}, \tag{6.12}$$

which converges to zero with increasing n for any fixed $\alpha < 1$. The condition $\alpha < 1$ is thus sufficient for reliability, with an implied reliable inference rate of α bits per symbol.

Proof. Let Δ_j be the error in the estimate $\hat{r}_j$ of r_j. Then, by appeal to the Chernoff bound derived in the note to Section 1, we have

$$P(\tilde{r}_j \neq r_j) = P(\Delta_j \geq 1/2) \leq \exp[-(8\mathrm{var}(\Delta_j))^{-1}] = \exp[-(8vD_j)^{-1}]$$

where D_j is the jj^{th} element of the matrix $(A^T A)^{-1}$. This then has the implications

$$P(\tilde{r} \neq r) \leq \sum_j \exp[-(8vD_j)^{-1}]$$

and

$$\overline{P}(\tilde{r} \neq r) \leq \alpha n E\{\exp[-8vD_j)^{-1}]\}, \tag{6.13}$$

where the expectation is over random variation of a, and we assume this such that the expectation is independent of j. Now, D_j^{-1} can be written

$$D_j^{-1} = \inf_\beta \sum_{k=1}^n \left(a_{jk} - \sum_{i \neq j} \beta_i a_{ik} \right)^2, \tag{6.14}$$

i.e. as the total squared error of the linear least square estimate of the elements of a_j in terms of the corresponding elements of a_i $(i \neq j)$. But it follows from the statistical theory of the linear model that, for random a_{jk}, expression (14) is distributed as P times a chi-squared variable with $n - p = n(1 - \alpha)$ degrees of freedom. With this we can evaluate the expectation in (13) and deduce inequality (12). $\square$

Of course, p cannot actually equal n, but the remarkable conclusion is that p can differ from n by a term of size $o(n)$, consistent with reliability. Our proof of course holds only for the Gaussian case, for which D_j has a known and simple distribution. However, one can expect analogous results for much more general cases.

6.6. Stability constraints for the quantised regression

Consider the recursive realisation (11) of the matrix inversion. We noted that it was necessary for stability of this recursion that the matrix $W_0^{-1}W_1$ should be stable. One might think this assured, since all the elements of this matrix are of probable order $n^{-1/2}$. However, we shall see that the fact that the matrix is of dimension $\mathcal{O}(n)$ can lead to instability unless α obeys the stronger bound

$$\alpha < \alpha_c = 3 - \sqrt{8} \doteq 0.172. \tag{6.15}$$

That is, the condition $\alpha < 1$ is both necessary and sufficient for reliability if the linear equation system (10) is solved exactly. The stronger condition (15) is associated with recursive solution (11) of the system, and is both necessary and sufficient for stability of this recursion.

We now see the reason for this bound on basis rate. We can solve equation (10) and achieve an exponentially large memory size with a vanishing error rate (12) if $A^T A$ is nonsingular in some uniform sense, and we shall see that this is assured (in the limit of large n) as long as $\alpha < 1$. However, we shall also see that the recursive implementation (11) of the algorithm fails if α equals or exceeds the critical value $3 - \sqrt{8} \doteq 0.172$.

The matrix $A^T A$ is non-negative definite, and so all its eigenvalues are real and non-negative. However, in order to treat the questions raised above we have to determine its asymptotic $(n \to \infty)$ eigenvalue distribution more explicitly. The classic Wigner semi-circle theorem gives this distribution in the case when A is square, and so $\alpha = 1$; the case $\alpha < 1$ has been treated by novel methods by Wachter (see Mallows and Wachter (1972) and Wachter (1974, 1978, 1980)).

Let λ be a random eigenvalue under the asymptotic distribution, so that $H(z) = E[(z - \lambda)^{-1}]$ is the Hilbert transform of this distribution.

Theorem 6.3. *Consider the matrix $U = n^{-1}A^T A$ under the assumptions that the elements a_{jk} of the matrix A are independent random variables, taking the values ± 1 with equal probability. Then the Hilbert transform of* *the asymptotic eigenvalue distribution of U has the evaluation*

$$H(z) = \frac{1}{2\alpha z}[z - 1 + \alpha - \sqrt{z^2 - 2(1 + \alpha)z + (1 - \alpha)^2}], \qquad (6.16)$$

where that branch of the function is taken for which the root is positive for large positive z.

A proof is given in the Appendix, in a version which makes clear that Theorem 6.3 will still hold if one assumes merely that the trace elements a_{jk} are independently and identically distributed with zero mean, unit variance and bounded moments. These conditions could certainly be relaxed even further.

Theorem 6.4. (i) *The asymptotic eigenvalue distribution of U has as support the interval* $(1 - \sqrt{\alpha})^2 \leq \lambda \leq (1 + \sqrt{\alpha})^2$, *with the addition of an atom of size* $(\alpha - 1)/\alpha$ *at* $\lambda = 0$ *when* $\alpha > 1$.

(ii) *In the limit* $n \to \infty$ *the matrix U is singular if and only if* $\alpha \geq 1$.

(iii) *In the limit* $n \to \infty$ *the iteration (11) fails to converge if and only if* $\alpha \geq \alpha_c = 3 - \sqrt{8} \doteq 0.172$.

Proof. (i) The points of the support will be the singularities of $H(z)$, located in the present case as the values of z which render the expression under the square root in (16) negative, plus a possible singularity at $z = 0$. The first set then forms an interval $\mathcal{I}$ with end-points $\lambda = (1 \pm \sqrt{\alpha})^2$, the values of z which render the square-root expression zero.

The selection of the branch will imply that we take the positive square root in (16) if $z > (1 + \sqrt{\alpha})^2$ and the negative square root if $z < (1 - \sqrt{\alpha})^2$. We thus find that expression (16) has no singularity at $z = 0$ if $\alpha < 1$ and a simple pole of residue $(\alpha - 1)/\alpha$ if $\alpha > 1$. This pole corresponds to the atom of probability asserted — a reflection of the fact that $p - n$ of the p eigenvalues of the matrix $U = n^{-1}A^T A$ will be zero if $p \geq n$.

(ii) As a increases, then the lower limit of the interval $\mathcal{I}$ reaches zero first when $\alpha = 1$. It is at this point that U first becomes singular, and stays so, as just argued. This is consistent with the conclusions of Theorem 6.2.

(iii) The matrix $W_0^{-1}W_1$ of (11) will cease to be a stability matrix when $\mathcal{I}$ first contains values outside the interval (0,2). This occurs first when $(1 + \sqrt{\alpha})^2 = 2$, giving the determination of α_c asserted. $\qquad \Box$

A network which realises the recursion (11) and the discretisation (8) is that depicted in Fig. 6.1. A further output stage $\tilde{y} = A\tilde{r}$ could be added if one wished to make the memory autoassociative, and this output be fed back to the input if one wished to achieve a memory store. As formulated,

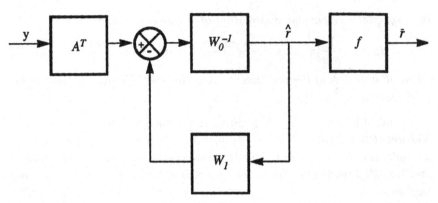

Fig. 6.1. A network which would realise the associative memory of recursion (11) and discretisation (8); a 'quantised regression'.

the memory stores p bits with a bound on the probability of error per pass given by (12). The network would have $\mathcal{O}(n)$ nodes but $\mathcal{O}(n^2)$ arcs and proportional power consumption. This latter could be reduced to $\mathcal{O}(n \log n)$ by adoption of a sparse coding (i.e. one in which only a fraction $\mathcal{O}(\log n / n)$ of the a_{jk} are non-zero).

An interesting point is that the critical basis rate α_c is independent of v. The reason for this is that the presence of observation noise does not affect the character of the matrix $A^T A$. Nor even does plant noise; i.e. noise which enters the recursion (11) additively. This will perturb the solution of (11), but will not affect its stochastic convergence. However, for the usual stochastic version of the Hopfield network one finds that the critical value α_c falls with increasing temperature (as measured by the variance of plant noise). The reason is that, in the Hopfield case, noise enters the actual autoassociative recursion. Interaction of noise with the nonlinearity of f introduces a 'multiplicative' effect, and the whole calculation then becomes temperature-sensitive.

6.7. The Hopfield net

Having dealt so extensively with the recognition of compound traces, we are now in a position to return to the Hopfield net, introduced in Section 5.4. We take the view that this not quite 'right' as it stands, but that it has a fascination for a particular reason. The reason is that the very problem which it in fact partially addresses has not yet been clearly perceived. We claim to identify it, in Chapter 8, as the problem of fading data.

Let us first discuss some properties of the network additional to those mentioned in Section 5.4. Performance of the net is usually evaluated under the random trace assumption: that the elements. a_{jk} of the p traces are independent random variables, taking values ± 1 with equal probability. Such trace statistics produce a definite lack of feature, but have some interest and lead to a manageable analysis. Performance characteristics quoted below are for these trace statistics, a hard threshold function, and noiseless operation of the net ('zero temperature'), unless otherwise stated.

Hopfield found in his paper (1982) that the ratio $\alpha = p/n$ was crucial, and that a critical value α_c of this ratio should not be exceeded if trace recognition was to be reliable. Early numerical experiments estimated this critical value as about 0.15. Theoretical analysis, by the replica method of statistical physics (Amit *et al.* (1985)) and by refined versions of more direct arguments (Amari and Maginu (1988)) produced the evaluation $\alpha_c \doteq 0.138$.

The implication is that the number of traces p cannot grow faster than linearly with n, if recognition is to be reliable. This seems to conflict inexplicably with the expectation of Section 1: that the number of traces should be able to grow exponentially with n. However, the analysis of Section 6 points to the reason for the condition. That analysis was concerned with the simpler case of linearly compounded (and rationally inferred) traces, but nevertheless implied the necessity of an upper bound $\alpha_c \doteq 0.172$ on α. This bound was not implied by reliability considerations, but by the necessity for stability of the recursion estimating the loading vector r. The Hopfield network is intrinsically recursive, and one can identify the constraint $\alpha < 0.138$ as a stability rather than a reliability condition.

Another feature of the Hopfield network is that, as well as equilibria corresponding to the p basic traces, it admits equilibria corresponding to compound versions of these p traces (Gardner (1986), Komlös and Paturi (1988)). These compounds cannot quite be linear superpositions — the restriction of the elements of y_t to the values ± 1 excludes that — but they are as close to being so as that constraint permits. They are termed 'spurious' in the literature and regarded as undesirable artefacts, to be suppressed by special measures. However, we would regard them rather as a sign of grace. The Hopfield network asserts its right to exponentially many stable equilibria precisely by the appearance of these unintended compound equilibria. Expressed rather more objectively, we shall see in Chapter 8 that the compound equilibria have virtually been designed into the net, and will not be eliminated by purely cosmetic measures. The degree of stability of these equilibria, i.e. the depth of the corresponding basins of attraction, decreases with increasing compounding, but that is as one would expect. Nor need the compound traces be physically meaningless — sensory images are in most cases compounds of more basic images, as we have emphasised.

If one tries to place the Hopfield net in the gallery of associative memories then there are so many points which can be made that they are best listed formally. On every point we shall try to give both the positive and the negative aspects.

(1) The Hopfield net seems to lack an obvious derivation which would establish its rationale. Lippmann (1987) makes the point that it must necessarily suffer by comparison with the Hamming net, which implements the statistical analysis which would be optimal under the assumptions stated in section 5.2. He further asserts that this difference in performance is borne out in tests on character recognition, pattern recognition and bibliographic retrieval.

(2) The Hamming net achieves its final inference in a single pass — one then asks why the Hopfield network may do so only after several iterations. To answer this, we note that the net is implicitly operating as a storage memory, storing (for a while) the data, possibly degrading, with which it was initially furnished, and storing (indefinitely) the evoked form of the trace to which it converges. It is also achieving an iterative version of the matrix inversion required for the quantised regression.

(3) The net will operate as a successful associative memory only if the traces are almost orthogonal (the independently generated random binary traces mentioned above are sufficiently orthogonal). If one sets it up with non-orthogonal traces then it can fail completely. For example, Kumar, Saini and Prakash (1996) report cases in which the net annihilates the memory of all encoded vectors; such cases can be found for p as small as 3 for any n greater than 5. (That is, even if y_0 is set equal to one of these traces y_t will diverge from it with increasing t.)

This is a serious defect. It is partially remedied by the proposal that the autoassociative operator should be modified from fAA^T to $fA(A^TA)^{-1}A^T$, although performance of the resulting net has defied analysis.

(4) The net can be derived from an extremal principle: that it achieves monotonic increase of the quadratic form y^TWy on the set of binary vectors (i.e. vectors with elements ± 1). As always, this is a powerful property. Because of it one can make strong convergence statements,

and a bridge can be established with the models and methods used by physicists to deal with spinglasses — randomly generated structures which nevertheless show collective effects (see the references to Amit *et al.*, Gardner, and Kamp and Hasler).

However, this particular extremal principle seems to be accidental — no significant meaning can be read into it. In Chapter 8 we appeal to an extremal principle which really has a statistical basis.

(5) There are the two important features which we have discussed above: the occurrence of 'spurious' equilibria and the need (and its explanation) for an upper bound on $\alpha = p/n$.

(6) The net seems to have an unnecessarily high dimension: n rather than the p of the Hamming net, Furthermore, even if the matrix A is sparse, the matrix AA^T is not, with the consequence that the Hopfield net has something like n^2 synaptic connections.

Against this, one can argue that operation in the full number of dimensions allows one to keep one's options open — for example, to add new memory traces if needed. In other words, plasticity is retained, in that aspects of the data which might prove important are not discarded prematurely.

A further point is that the Hopfield net operates successfully even if the activation function f is soft. In the language of the physical spinglass analogue, the net can perform even at a positive temperature (since randomness of operation can to some extent be seen as a softening of the activation function). However, the combination of dynamic operation and the high dimension of the net means that collective effects can still be realised. Specifically, effective separation of basins of attraction can still be achieved.

(7) The data y are fed into the net as an initial setting of the network state rather than as a driving input. This has the effect that the net, when used as a storage memory, can only be reset by a bodily change of its whole state.

However, this may be an inescapable feature if data cannot be stored without degradation; see Chapter 8.

(8) It is notable that the activation function f is applied at every node of the network —no 'linear' nodes are allowed, for example. However, this would be a natural constraint if the net were indeed to be realised as a neural network (either natural or artificial), and the constraint also enforces the physically necessary limitation on local signal strength. It

is also worth repeating the point made in (6): that the net will then perform even if this activation function is soft.

(9) Finally, there is the point that the synaptic matrix $W \propto AA^T$ of the Hopfield net can be 'learned' by the so-called Hebbian rule over repeated trials. Suppose that the data vector y has the form $y = a + \epsilon$, where a takes one of the p trace values randomly and equiprobably and ϵ is an error vector, independent of a and with $E[\epsilon\epsilon^T] = V$. We would then have

$$V_{yy} = E[yy^T] = p^{-1}AA^T + V$$

(where the expectation is taken also over random occurrences of the given traces). The synaptic matrix W chosen for the Hopfield net is thus proportional to that part of V_{yy} which is attributable to statistical variation of the trace presented. This is often taken as an instance of 'Hebbian learning' — a point which we discuss further in Chapter 10.

Chapter 7

Preserving Plasticity: A Bayesian Approach

The associative memories or 'recognisers' constructed in Chapters 5 and 6 were based upon the classical formulation of statistical inference; in particular, upon the method of maximum likelihood. It is argued in Section 1 that a Bayesian formulation is closer in spirit to the 'learning from experience' which is so characteristic of the neural approach. This is a viewpoint which leads to interesting conclusions, although it is in Chapter 8 that we finally settle on the approach which, in the author's view, poses and answers the right questions.

The classical and Bayesian approaches are not mutually inconsistent, but there is a shift in emphasis which affects the mathematics interestingly. We shall see that the Bayesian approach leads to dynamic equations which are somewhat of a Hopfield character, and that these are in the full n-dimensional variable rather than in a lower-dimensional parametric variable r.

However, the dynamics derived are not completely Hopfield in character. There is indeed a sigmoid activation function, which appears as the consequence of a robustness demand. However, the data vector is injected as a driving input rather than an initial value. Furthermore, the feedback turns out to be negative rather than positive in character, for reasons explained in Section 2.

7.1. A Bayesian view

In Chapters 5 and 6 we supposed that the memory traces a_j were known, and that it was then a matter of seeking the linear function $\sum_j r_j a_j$ of these traces a_j which best matched the current observation vector y. This view is the classical statistical one, of regarding the coefficients r_j as parameters. However, if one thinks of how learning actually takes place, it seems more likely that r and y should both be regarded as random variables; that one builds up experience of their joint statistical variation over the course of time and seeks, on the basis of this experience, to determine the best predictor of r in terms of y in a given case.

For example, imagine an animal observing other animals of the same or other species which are significant to him, either socially, as danger or as quarry. He may accumulate both visual and olfactory information, say. We can think of r as being a condensed version of the visual evidence — a rough indication of species present and their numbers — and of y as being the olfactory evidence. It is observation of both aspects which provides 'supervised learning'. In cases when the animal receives only the scent then he would like to determine its cause; to reconstruct the now missing visual data (the types and numbers of animals which left the scent) as well as possible. In other words: to estimate r from y.

One could indeed imagine that the animal had formed memory traces, in that he had stored in his memory the typical scent of a deer, say, or the typical scent of a receptive female of his own species. However, a more primitive stage in development, and a more adaptable one, is that of simply forming a statistical association of stimulus y with cause r. Since y and r are both regarded as random variables, this is a Bayesian approach.

Let us take the usual starting case, and assume that y and r are jointly normally distributed. We can also assume, without further loss of generality, that these variables have been reduced to zero mean. The covariance matrix of r and y, for example, is then $V_{ry} = E(ry^T)$.

Then the obvious estimate of r in terms of y is

$$\hat{r} = E(r|y) = V_{ry}V_{yy}^{-1}y. \tag{7.1}$$

We know from Chapter 2 that this is the projection estimate, having a number of other optimality interpretations: it is the conditionally most probable value and it is also the linear estimate of minimal mean-square error.

The estimate is necessarily linear in this case, but what is interesting is that it involves inversion of the $n \times n$ matrix V_{yy} (rather than, as in (6.5), of the $p \times p$ matrix $A^T A$). This is indeed reminiscent of the Hopfield net, for which the synaptic matrix is closely related to V_{yy}. There is, of course, no nonlinearity in relation (1), because of our assumption of normality and of the fact that we have not yet considered neural implementation of the calculation. Nevertheless, the occurrence of the inverse V_{yy}^{-1} makes it likely that some variant of the Hopfield net can be seen as motivated in a Bayesian framework.

Note also the conservation of plasticity. The principal step in (1) is the calculation of the n-vector

$$\lambda = V_{yy}^{-1} y, \tag{7.2}$$

a calculation independent of the choice of the variable r which is to be estimated. Once one has decided on the relevant variable then its estimate is yielded in terms of λ by

$$\hat{r} = V_{ry} \lambda. \tag{7.3}$$

Matters are very different for the parametric case of Chapter 6. If we had wished to change the definition of r (by addition of new traces, for example), then the inversion $(A^T A)^{-1}$ would have had to be recalculated.

However, there is a difficulty. The matrix V_{yy} is of order n, which we expect to be large, and so inversion is laborious. There is also the danger that the matrix may be ill-conditioned, in that it is almost singular — a concern likely to become more acute as n increases. Of course, even if V_{yy} is singular, the combination $V_{ry} V_{yy}^{-1}$ is still meaningful. Expression (2) is also meaningful, provided that y lies totally in the orthogonal complement of the null space of V_{yy} (as it should do, if it is indeed a sample of a random variable with covariance matrix V_{yy}). However, the presence of rounding errors, for example, could be enough to render the calculation quite impracticable.

It is helpful to take a Bayesian form of the model

$$y = Ar + \epsilon$$

of Chapter 6, assuming the p-vector r and the n-vector ϵ normally and independently distributed with zero means (after a standardisation) and covariance matrices U and V respectively. Then $V_{rr} = U$, $V_{ry} = UA^T$ and $V_{yy} = AUA^T + V$. The alternative characterisation of $\hat{r}$ is that it is the value of r minimising the quadratic form

$$Q(r, y) = \frac{1}{2} r^T U^{-1} r + \frac{1}{2} (y - Ar)^T V^{-1} (y - Ar) \tag{7.4}$$

which appears in the exponent of the joint Gaussian density. The stationarity condition can be written

$$r = UA^T V^{-1} (y - Ar), \tag{7.5}$$

which is in fact consistent with (1). It also shows that the λ of (3) can be identified with $V^{-1}\hat{\epsilon}$, where $\hat{\epsilon}$ is the estimated noise vector $y - A\hat{r}$.

It is likely that equation (5) could be made the basis of an iterative estimation procedure. However, let us first consider a more realistic model.

7.2. A robust estimation method

The danger of deducing procedures by optimisation on the basis of a particular model is that they may have no validity if reality departs even slightly from that model. Procedures which do not show this sensitivity are *robust*.

The estimate (1) of the last section showed a lack of robustness in that it would be nonsensical if the observation vector y had even the slightest component in the null space of its assumed covariance matrix V_{yy}. In the case of our hypothetical animal, the corresponding event for him would be to encounter a scent which was simply inexplicable in terms of his experience.

This would be automatically guarded against if the noise ϵ were statistically non-degenerate, in that its covariance matrix V were nonsingular. The degree of robustness which this induces in the estimate would be greatly increased if we assumed more: that, not merely does the noise vector ϵ have a non-degenerate distribution, but also that this distribution is heavy-tailed (platykurtic). That is, that it allows a higher proportion of large deviations from expectation than would a normal distribution of the same variance.

Let us suppose, as before, that r is normally distributed with mean zero and covariance matrix U, so that its probability density is proportional to $\exp[-\frac{1}{2}r^T U^{-1} r]$. Suppose that the noise vector ϵ is distributed independently of r, and that its elements ϵ_j are distributed independently and identically with probability density proportional to $\exp[-D(\epsilon_j)]$. The form Q of (4) then becomes

$$Q(r, y) = \frac{1}{2} r^T U^{-1} r + \sum_j D[(y - Ar)_j], \qquad (7.6)$$

and to estimate r as the value minimising this expression is equivalent to estimating state x as the value maximising $P(x, y) = P(x)P(y|x)$.

Let us build in the heavy-tail property by assuming that the ϵ_j distribution has exponential tails rather than Gaussian tails. That is, that

$D(\zeta)$ grows as $|\zeta|$ rather than as ζ^2 for large $|\zeta|$. This implies that the derivative

$$f(\zeta) = \frac{dD(\zeta)}{d\zeta}$$

will tend to positive and negative asymptotes respectively as ζ tends to $\pm\infty$, at least under smoothness assumptions. If we assume also that the ϵ_j distribution is unimodal then $f(\zeta)$ will in fact be a sigmoid function. The sigmoid activation function thus makes an entry as a consequence of platykurtosis and as a champion of robustness.

Minimising expression (6), we find that the estimate $\hat{r}$ should be a root of the equation

$$r = UA^T f(y - Ar), \tag{7.7}$$

a generalisation of (5).

Note that equation (7) has a solution, and that this solution is unique. Expression (6) is continuously differentiable as a function of r, and tends to $+\infty$ as $|r| \to \infty$. It thus has a minimum, and so at least one stationary point. The stationarity condition (7) thus has at least one solution. But expression (6) is strictly convex as a function of r, and so this stationary point is unique.

It is helpful to recast equation (7). The calculation is immediate, but the conclusion significant enough to deserve emphasis.

Theorem 7.1. *Under the distributional assumption* (6) *the estimated residual* $\hat{\epsilon} = y - A\hat{r}$ *is the unique solution of the equation in* ϵ

$$\epsilon + Wf(\epsilon) = y, \tag{7.8}$$

where W is the $n \times n$ matrix

$$W = AUA^T. \tag{7.9}$$

Proof. Equation (8) follows immediately from (7), and uniqueness of solution from monotonicity in ϵ of the left-hand member. □

W is just the synaptic matrix of the Hopfield net (at least if we assume U proportional to the identity): the non-noise component of V_{yy}. Equation (8) is an equation in the full n-vector, and it is close to being a version of the Hopfield equation. In terms of the transformed variable $\eta = f(\epsilon)$ it can be written

$$\eta = f(y - W\eta) \qquad (7.10)$$

which might be given the recursive form

$$\eta_{t+1} = f(y - W\eta_t). \qquad (7.11)$$

Relation (11) has a distinct Hopfield look about it, but it differs from the Hopfield recursion in two essential respects. One is that the observation vector y appears as a driving input rather than as an initial value. That is, there is an implicit assumption that the data can be stored, either inside the net or (as a sustained stimulus) outside it.

The other point is that the feedback term $W\eta$ appears negatively in the argument of f, rather than positively, as it did in the Hopfield equation. That is, it appears as a correction to the raw observation vector y rather than as itself some idealised version of y. This is understandable if the original data vector remains available. Alternatively, the negative sign can be seen as a consequence of the positive loading of the two terms in ϵ in (8). This representation of y corresponds to a division of y between the two sources of variation: prior variation of r and sampling variation of ϵ. A deviation of y from zero represents a degree of deviation from expectation, and the improbability implicit in this deviation must be divided between the two sources of variation.

7.3. Dynamic and neural versions of the algorithm

In what follows we shall continue the dangerous but convenient convention already established in equations (7) and (8): of using r and ϵ to denote estimates (provisional or final) of the actual variables r and ϵ.

A steepest-descent equation

$$\dot{r} = -\kappa \frac{\partial Q(r, y)}{\partial r}$$

would yield a dynamic version of the estimating relation (7), whose solution would certainly converge to the value of r minimising Q. However, if we wish to 'preserve plasticity', i.e. avoid parametric assumptions, then it is equation (8) that we should solve, with W deduced from a learning rule rather than calculated from a prescribed A.

A dynamic version of (8) would be

$$\dot{\epsilon} = \kappa[y - \epsilon - Wf(\epsilon)]. \tag{7.12}$$

However, a neural version of this relation would rather be of the form

$$\mathcal{L}\epsilon = d^2[y - Wf(\epsilon)]. \tag{7.13}$$

Here $\mathcal{L}$ is the second-order differential operator introduced in Section 3.4, which will be an essential component of the more realistic neural models to be adopted in Part III. We have supposed that it has the form $L(\mathcal{D}) = \mathcal{D}^2 + c\mathcal{D} + d^2$, where $\mathcal{D} = d/dt$; the term $d^2\epsilon$ thus takes up the linear term in ϵ in (12). The operator is known to be stable, in that the zeros of the polynomial have negative real part; this implies positivity of both c and d^2.

Relation (13) is not a steepest-descent equation itself, but is near enough to one to ensure the desired limit properties.

Theorem 7.2. *The solution $\epsilon(t)$ of (13) converges to $y - Ar$, where r is the value minimising $Q(r, y)$.*

Proof. It is assumed that the parametric model holds. Let us write $\epsilon = y - Ar + \chi$, where r is to be determined and χ is an n-vector lying in the null space of A; i.e. orthogonal to all the columns of A. Then relation (13) implies that

$$\mathcal{L}\chi = 0, \quad \mathcal{L}r = -d^2 U A^T f(y - Ar + \chi).$$

The first equation implies that $\chi \to 0$ with increasing time, because of the stable character of $\mathcal{L}$, Setting $\chi = 0$ in the second relation, we obtain an equation which can be written

$$U^{-1}(\ddot{r} + c\dot{r}) = -d^2 \frac{\partial Q}{\partial r}. \tag{7.14}$$

Multiplying by $\dot{r}^T$ and integrating, we obtain the relation

$$\frac{1}{2}R(t) + d^2 Q(r(t), y) = k - c \int^t R(\tau)d\tau, \tag{7.15}$$

where k is a constant of integration and R is the positive definite quadratic form $\dot{r}^T U^{-1} \dot{r}$, evaluated at the time indicated. The left-hand member of equation (15) is bounded below; the same can be true of the right-hand member only if $R(t)$ tends to zero with increasing t. This implies that $\dot{r}$ converges to zero, as must then $\partial Q/\partial r$, in virtue of (14). We have thus proved that χ tends to zero and r to the value which minimises Q. $\square$

Chapter 8

The Key Task: the Fixing of Fading Data.
Conclusions I

8.1. Fading data, and the need for quantisation

In deducing inference rules we have assumed that the data vector y can be held and fed into the calculations as they progress, either because the sensory input is sustained or because, once received, it can be stored in the system. However, suppose that it is not sustained, and that it is subject to continued degradation if held in the system — that is what we mean by 'fading data'. After all, if continuing noise in the system means that the memory of an evoked trace degrades, then the memory of stored data will equally degrade, and stabilising measures must be taken on both counts.

In such cases the data must be regarded as part of the prescription of the initial state of the dynamic system, rather than as a continuing input to the system. This is indeed assumed for the Hopfield network: the 'idealised data vector' is the state variable of the system; it is set equal to the actual data vector y initially, and the dynamics of the system act to modify it to something like what y would have been had the most plausible pure trace been observed. That this course is followed is an indication that more than simple inference is at stake; the inference must be conducted in a race against fading of the data.

The question is now whether there is a principle which will provide a general rationale for such situations: situations in which the limited computational resources available must be devoted, not merely to inference, but also to the protection of the data from degradation while this inference is being performed. One suspects that an ideal solution will resemble the Hopfield network in some respects, but not in all.

We shall now use y to denote an 'idealised data vector', a dynamic variable $y(t)$ of the system whose initial value $y(0)$ is just the actual data vector. One thing that is clear is that y must be quantised, in that it must be brought to one of a number of discrete points of attraction as soon as possible if drift is to be arrested. To take the simplest possible case, suppose that y is a scalar obeying the dynamic equation

$$\dot{y} = g(y) + \eta. \tag{8.1}$$

Here $g(y)$ is a function which specifies the dynamics, and which can be chosen subject to physical constraints (e.g. that g is bounded). The term η is continuing process noise, which design cannot affect and which is the source of data degradation. We can, for definiteness, think of it as white noise of power c.

Suppose that we avoid deliberate distortion of the data, by taking $g \equiv 0$. Then y follows a random walk, and drifts indefinitely. Suppose, on the other hand, that we take g as a function with a number of zeros, as in Fig. 8.1, those zeros at which g has negative derivative then constituting stable equilibria of the noiseless system. In the noisy case, the y variable will find its way to the neighbourhood of one of these equilibria. The probability that the equilibrium first found is not the one in whose basin of attraction the initial value y lay will be of order $\exp(-k_1/c)$ for some positive k_1, dependent on $y(0)$ but not on c. The expected time for which it stays in this basin of attraction will be of order $\exp(k_2/c)$ for some positive k_2, dependent on the depth and breadth of the basin of attraction, but not on c.

So, the introduction of this 'quantisation' introduces a degree of distortion, in that y is quickly confined to the neighbourhood of one of these equilibrium values. On the other, it effectively arrests drift over a horizon whose logarithm is of order c^{-1}. There is a trade-off between the two effects: if one decreases distortion by taking a denser set of equilibria

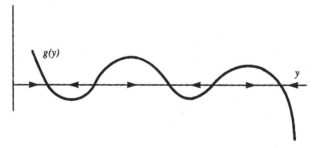

Fig. 8.1. A function g which realises a quantisation, in the scalar version of equation (3.1).

then one decreases the depth of the corresponding basins of attraction, from which escape then becomes more frequent.

Suppose one has a parametric model for the data; for example, the familiar representation,

$$y = Ar + \epsilon, \tag{8.2}$$

of an initial data n-vector y as a linear function of the parameter vector r plus observation noise. One might think that the sensible method of quantisation would be to derive an estimate $\hat{r}$ of r and then to quantise this suitably, as we proposed in Chapter 6. However, quantisation has to be achieved early, *while* the inference is being performed. Indeed, the inference step may be implicit, in that y is led to the equilibrium value close to $A\tilde{r}$, where $\tilde{r}$ is the most plausible quantised value of r, without $\hat{r}$'s ever being calculated. This is autoassociation with the inference step suppressed.

8.2. The probability-maximising algorithm (PMA)

If we are to be guided by some kind of optimality consideration then we must take a model, at least provisionally. Let us then indeed take the linear model (2), assuming that the elements of the observation noise vector ϵ are normally and independently distributed with zero mean and variance v. More general statistics are easily handled, but it is best to keep the treatment as transparent as possible in a first approach. We shall assume that the vector parameter r has a prior distribution with measure μ. The probability density function of the initial data vector is then

$$\rho(y) = (2\pi v)^{-n/2} \int \exp\left[-\frac{1}{2v}|y - Ar|^2\right] \mu(dr). \tag{8.3}$$

The quantisation of which we spoke is to be achieved by taking the prior distribution μ as a discrete one, confining r to a discrete set $\mathcal{R}$ and giving a value r of this set the prior probability π_r. This discrete distribution is not regarded as constituting the true prior, but rather as an acceptable quantisation of that prior. Under this supposed prior expression (3) takes the form

$$\rho(y) = (2\pi v)^{-n/2} \sum_{r \in \mathcal{R}} \pi_r \exp\left[-\frac{1}{2v}|y - Ar|^2\right]. \tag{8.4}$$

To avoid degeneracy we suppose all r-values distinct and all π_r strictly positive.

We shall specify the data-preserving dynamics by letting y change in such a way as to increase $\rho(y)$ — in fact, to follow a steepest ascent path for this function. The expectation is then that y will converge to a value close to $A\tilde{r}$, where $\tilde{r}$ is the value in $\mathcal{R}$ which seems most probable conditional on the initial value of y.

This approach is something like a dual to maximum likelihood. Likelihood maximisation would infer the estimate as the permissible r-value maximising $\rho(y|r)$. In the present case we choose rather to let y seek the local maximum of $\rho(y)$ which is found by steepest ascent from the initial value $y(0)$. Equivalently, we can regard this as a steepest descent rule for the potential function $-\log \rho(y)$. The actual dynamic equation corresponding to the noiseless version of (1) is then

$$\dot{y} = \kappa v \frac{\partial \log \rho(y)}{\partial y}, \tag{8.5}$$

where κ is a suitable constant. The v is inserted for convenience. We shall also find it convenient to define the weights

$$\phi_r = \pi_r \exp\left[-\frac{|Ar|^2}{2v}\right], \quad (r \in \mathcal{R}). \tag{8.6}$$

Theorem 8.1. *The steepest descent equation for the potential function* $-\log \rho(y)$*, with* $\rho(y)$ *given the quantised form* (4)*, is*

$$\dot{y} = \kappa[AF(A^T y) - y], \tag{8.7}$$

where

$$F(x) = \frac{\sum_r r\phi_r e^{x^T r/v}}{\sum_r \phi_r e^{x^T r/v}}. \tag{8.8}$$

This can equivalently be written

$$\dot{y} = \kappa[AE(r\,|\,y) - y], \tag{8.9}$$

where the conditional expectation is that based on the joint (r, y) *distribution implied in expression* (4).

Proof is by direct verification; one sees from (8) that indeed

$$F(A^T y) = E(r\,|\,y), \tag{8.10}$$

where the expectation is calculated on the assumption that the discrete r-prior holds and y is related to r by (2).

It is equation (7) which is our suggestion for solution of the fading-data problem; we shall term it the *probability-maximising algorithm*, abbreviated to PMA. It will of course be subject to a disturbing noise input as in (1), but its bowls of attraction centred on the points $y = Ar$ ($r \in \mathcal{R}$) should proof it against that. The principle was suggested rather than derived, but it does seem to give natural dynamic expression to the reconciliation of data-fidelity and quantisation. We shall demonstrate in the next chapter that it has optimality properties. The 'potential function' $- \log \rho(y)$ plainly has more motivation than that of the Hopfield net: $-y^T W y$ on $\{-1, 1\}^n$.

The form of equation (7) already suggests a network realisation. We shall see in the next section that the function $F(x)$ is indeed a vector version of a generalised sigmoid function. The PMA (7) has then just the character which the discussion of Section 1 would have led us to expect. It incorporates the nonlinear positive feedback term $AFA^T y$ which both reinforces the image y and reshapes it in the direction of a quantised version. It also incorporates a linear negative feedback term which assures stability.

The operator AFA^T is essentially the continuous-time autoassociative operator. Comparison of this with the operators AMA^T or AfA^T of the Hamming net (see Chapter 5) suggests that the probability-maximising principle implies Hamming rules rather than Hopfield rules. We shall see that the PMA indeed acts as a dynamic and autoassociative version of the simple Hamming net, although with some special characteristics.

In terms of the variable $x = A^T y$ the PMA takes the reduced form

$$\dot{x} = \kappa [A^T A F(x) - x]. \tag{8.11}$$

One would retain the full form (7) if one wanted the full autoassociative memory. However, the reduced form (11) is sufficient for estimation, since we see from (10) that $F(x)$ can be identified as the running estimate $E(r|y)$ of r.

The form of (11) suggests comparison with the classic estimation of r by minimisation of $|y - Ar|^2$, and the corresponding steepest descent equation

$$\dot{r} \propto A^T [y - Ar]. \tag{8.12}$$

However, in (12) it is r which is adapting to a fixed y, and in (7) it is y and r which are adapting simultaneously to achieve consistency with a relation $y = Ar$, where r takes a value close to the value in $\mathcal{R}$ which is most consistent with the initial y-value. In this sense the technique does indeed appear as the dual of maximum likelihood.

The effect of the PMA is that, starting from the initial data value, the variable y slides into the nearest potential well associated with a local maximum of $\rho(y)$. If v is small and the values Ar $(r \in \mathcal{R})$ are distinct then there is such a well for every r-value in $\mathcal{R}$. However, the wells widen as v increases, in an attempt to ensure that the initial y lies in the correct basin of attraction, despite the statistical increase in observation error ϵ. The effect of this will be that some basins merge and that the quantisation must be coarser — a consequence for performance and inferential capacity that one would have expected.

Actually, the 'precision' parameter is n/v rather than $1/v$, since we may expect $|y - Ar|^2$ to be of order n, whether r is correctly specified or not.

The effect of process noise η will be to perturb dynamics to the point that y may leave its current basin of attraction. However, as indicated above, the expected time taken for it to do so is exponentially large in the reciprocal of noise power.

8.3. Properties of the vector activation function $F(z)$

Relation (7) can be implemented on a net of n nodes and relation (11) on a net of p nodes, but the nonlinear operation F which is required is not of the simple form f to which we are used. We shall see in Section 5 that simplicity can be restored on an augmented net. However, there is some point to investigating the nature of F and possible immediate simplifications first.

A couple of properties follow immediately from the definition (8).

Theorem 8.2. (i) *The vector function $F(x)$ is non-decreasing in the sense that the $n \times n$ matrix of derivatives $(\partial F_j / \partial x_k)$ is non-negative definite.* (ii) *For a given p-vector w, let $r(w)$ be the value of r in $\mathcal{R}$ which maximises $w^T r$. Then $F(x)$ converges to the limit value $r(w)$ as x increases indefinitely in the direction of w.*

Proof. Let us denote a sum $\sum_r h(r)\phi_r \exp(x^T r/v)$ simply by $Sh(r)$. Then we have

$$\frac{\partial F(x)}{\partial x} = \frac{Srr^T}{S1} - \frac{(Sr)(Sr^T)}{(S1)^2} \tag{8.13}$$

which is a non-negative definite matrix. In fact, the expression on the right of equation (13) is just the covariance matrix of the random variable r under the 'posterior' distribution $\pi_r^* \propto \phi_r \exp(x^T r/v)$, $(r \in \mathcal{R})$.

Assertion (ii) follows immediately from expression (8) for $F(x)$.   □

So, assertion (i) implies that $F(x)$ is non-decreasing in the sense expressed there, and assertion (ii) implies that $F(x)$ has a finite limit as x tends to infinity in any given direction. These are consistent with sigmoid behaviour of the elements $F_j(x)$ of $F(x)$, but are by no means sufficient to ensure the full sigmoid property. To see this and rather more, consider expression (4) for the probability density $\rho(y)$ of y with a discretised prior. Suppose that $\mathcal{R}$ contains m elements and, to avoid degeneracy, suppose the values $Ar(r \in \mathcal{R})$ distinct. Then for v small enough $\rho(y)$ will have exactly m local maxima, these being at values $y_r = Ar + o(1)$ for small v, and the PMA equation (7) will have stable equilibria at these values.

Consider in particular the scalar case $p = 1$, when $F(x)$ would be a simple sigmoid if the neural pattern were followed. We can assume that $A = 1$ without loss of generality. We know then, from the argument above, that if v is sufficiently small, the equation $F(x) = x$ has m stable roots (i.e. roots at which $dF/dx < 1$); denote these by x_r. The situation must then be as Fig. 8.2: that the equation has $2m - 1$ roots, alternately stable and unstable, and that $F(x)$ is a simple sigmoid only in the case $m = 2$.

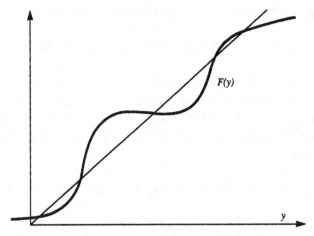

$F(y)$

Fig. 8.2.   The generalised sigmoid which the elements $F_j(x)$ of $F(x)$ would in general present.

Similar considerations apply in the vector case $p > 1$, which we shall follow up in the next section. Let us merely note for the moment that the deviation of y_r from $Ar(r \in \mathcal{R})$ for small v is due to the contribution to $\rho(y)$ from adjacent r-values in $\mathcal{R}$. As v increases from zero this effect becomes more pronounced, the local maxima of expression (4) will become weaker, and will progressively disappear. This corresponds to the merging of basins of attraction discussed in the last section.

8.4. Some special cases

We take it that the PMA provides a well-founded solution of the problem which the Hopfield net was effectively intended to address: to achieve autoassociative memory by steering the degrading data vector y quickly to the neighbourhood of the most compatible memory trace. This expectation will be justified in the next chapter. The next question is that of network implementation. The relatively complicated nature of $F(x)$ would seem to make realisation difficult, but there are at least two cases which demonstrate a simplification.

In order to achieve a neural realisation on the obvious net we should like the jth component $F_j(z)$ of $F(z)$ to be a simple sigmoid function $f_j(z_j)$ of the jth component z_j of z alone. There will indeed be such a simplification if the weights ϕ_r factorise into terms dependent upon the individual components r_j of the vector r $(r \in \mathcal{R})$:

$$\phi_r = \prod_{j=1}^{p} \phi_j(r_j), \quad (r \in \mathcal{R}) \tag{8.14}$$

One finds then immediately that the required reduction

$$F_j(x) = f_j(x_j) := \frac{\sum_r r_j \phi_j(r_j) e^{r_j r_j/v}}{\sum_r \phi_j(r_j)^{x_j r_j/v}} \tag{8.15}$$

follows. Furthermore, f_j will be a simple sigmoid if for every j the elements r_j of r can take only two values as r varies. That is, each basic trace can be present only at two intensities ('on' or 'off', presumably), the intensities of different traces being 'independent' on the basis of the modified prior ϕ_r.

With these assumptions AFA^r reduces to AfA^T and the PMA equation (7) to a recognisably 'neural' form. However, the assumptions seem in the highest degree Draconian. Even if the π_r factorised as indicated, the trace-dependent factor in the definition (6) of ϕ_r would not do so unless the traces were mutually orthogonal.

However, these observations do reveal one interesting point. The reduction of the vector sigmoid $F(x)$ to a vector of simple sigmoids (15) is achieved only by admitting all the 2^p compound stimuli which can be obtained from p individual stimuli, each with two possible levels. There will be a local maximum of $\rho(y)$ for each of these 2^p combinations if v is small enough. That is, there will be a full array of 'spurious' equilibria, and this is so *precisely because* the sigmoid responses were required to be local and simple. The Hopfield net, with its symmetric connection pattern and identical activation functions at every node, is as if designed to operate factorially — to admit all combinations of basic traces as traces in their own right.

In the literature the Hopfield net has been doctored in every possible way to try and eliminate spurious equilibria. These have included the admission of other dependences (than $W \propto AA^T$) of the synaptic weights w_{jk} on the memory traces (van Hemmen and Kühn (1988)), the admission of asymmetric weights (Xu, Hu and Kwong (1996), Parisi (1986), Nishimori, Nakamura and Shiino (1990) and Shiino (1990)), the admission of non-monotonic activation functions (Nishimori and Opris (1993), Morita (1993) and Yoshizawa, Morita and Amari (1993)) and the reduction of neuron gain (Waugh, Marcus and Westervelt (1991)).

However, none of these measures seem to be fundamentally motivated or particularly effective. If 'spurious equilibria' are regarded with disfavour, then this must be because multiple stimuli are assumed not to occur. If one took this view, then in the PMA algorithm one would then choose the prior distribution so as to admit only simple traces a_j. Expression (4) for $\rho(y)$ would then reduce to the form

$$\rho(y) = (2\pi v)^{-n/2} \sum_{j=1}^{m} \pi_j \exp\left[-\frac{1}{2v}|y - a_j|^2\right],$$

where π_j is the prior probability of trace a_j. The elements of $F(x)$ then take the form

$$F_j(x) = \theta \phi_j e^{x_j/v},$$

where θ is a common scaling factor

$$\theta = \left(\sum_k \phi_k e^{x_k/v}\right)^{-1}$$

and $\phi_j = \pi_j \exp[-|a_j|^2/(2v)]$.

This is indeed just a form of the Hamming net. The PMA equation (8.7) simplifies to

$$\dot{y} = \kappa(\theta A \phi f A^T y - y), \qquad (8.16)$$

where ϕ is the diagonal matrix with elements ϕ_j and the activation function f (acting on all the elements of the vector following it) is just the exponential function: $f(\xi) = e^{\xi/v}$. If v is small then the vector $\theta \phi f A^T y$ is virtually $M[A^T y - v(\log \phi_j)]$, the p-vector with a unit in the place corresponding to the best matching trace, and zeros elsewhere. The vector $\theta A \phi f A^T y$ is then virtually the best-matching trace itself. The net thus behaves as a dynamic and autoassociative version of the Hamming net, in that it swings y towards the best-matching trace. We have been led to incorporate the slight generalisation that prior probability π_j and trace strength $|a_j|^2$ may vary with j, and the selection effect is also somewhat fuzzy if v/n is not small. However, we recognisably have the simple Hamming net.

Of course, the exponential function is not sigmoidal, in that it is not bounded above. The bounding in the algorithm is supplied by the presence of the general scaling factor θ, which we could now write compactly as

$$\theta = (1^T \phi f A^T y)^{-1}. \qquad (8.17)$$

Relations (16) and (17) between them constitute the algorithm. We have already seen it partially proposed in (3.22); now we see it derived from the probability-maximising principle. It has an obvious network realisation, with f an arbitrary activation function which is operated far enough below saturation that it is sufficiently exponential. The deflation operation θ has to be realised by a subsidiary feedback loop; see the next section. This operation corresponds to the global inhibition which is widely recognised as being necessary if the net is to act selectively.

8.5. The network realisation of the full PMA

The stripping-down of the vector activation function $F(x)$ to the familiar vector of scalar activation functions $f(x)$ seems like a practical necessity, but the loss of so much structure must surely mean a loss of performance, when this is structure recommended by the probability-maximising principle. It is possible that one could synthesise $F(x)$ by a subsidiary McCulloch–Pitts net, but one is initially deterred from doing so by size considerations.

In general at least as many nodes are needed as there are terms in expression
(4), i.e. as there are values of r with positive prior probability. If we denote
this number by m, then m can well be exponential in p, the dimension of r.
The size of the net would thus be exponential in the size of the problem,
the very explosion which we have tried to avoid by linear combination of
the basic traces.

However, our later consideration of the neurophysiology of olfaction
('smell') in Chapter 14 encourages us to return to the idea. Smell is one
of the simpler of the senses, but the olfactory neuro-system seems more
complex, multiplexed and multi-staged than could be accounted for by a
simple associative memory, even if one takes into account all the additional
regulatory mechanisms which will be built into an organism. Moreover,
while odours can clearly be combined additively, the nose does not in fact
seem able to recognise mixtures as such, but only those mixtures which
have been learned. This is true also for the PMA: only those r-values are
recognised which have been assigned positive prior probability. This at least
means acceptance of the idea that not all combinations of basic odours will
occur, so that m may grow more slowly than exponentially with p.

Define the $p \times m$ matrix R whose columns are the m values of r allowed
in the prior distribution. Let us, as in the last section, write the exponential
function $e^{\xi/v}$ simply as $f(\xi)$, and understand that f applied to a vector acts
elementwise. We take ϕ as being the diagonal matrix with elements ϕ_r, now
of size m.

Theorem 8.3. *The PMA (7) can be written*

$$\dot{y} = \kappa(\theta AR\phi f R^T A^T y - y) \qquad (8.18)$$

where θ is the scaling factor

$$\theta = (1^T \phi f R^T A^T y)^{-1}. \qquad (8.19)$$

Proof is immediate, and we emphasise the result as a theorem because it
marks a significant point on the journey rather than an arduous one.

Equations (18) and (19) are of disconcerting simplicity. They are
nothing but equations (16) and (17) with the matrix of traces A replaced
by AR. In other words, we are again back to the Hamming net (in the
dynamic, autoassociative and somewhat blurred form to which the PMA
leads us), with just the particular feature that the m traces must all be
generated with the columns of A as basis.

This may seem like the snake of a game of snakes and ladders, in that we virtually find ourselves dumped back where we started in Chapter 5: with the Hamming net. However, that the PMA has such a simple interpretation is in fact a strength. We have indeed the Hamming net, but with two new features. One is the dynamic and autoassociative form in which it now appears. The other is the factorisation AR of the trace matrix, the two factors specifying respectively the basic traces and the combinations of these traces which are permitted. This factorisation has structural significance and surprisingly extensive implications.

The best way to see these implications is to work through a sequence of versions. Consider first the static form of relation (18); the set of equations

$$y = \theta AR\phi f R^T A^T y. \tag{8.20}$$

This relation, a loop in that it expresses y in terms of itself, is represented in the block diagram of Fig. 8.3. Note that there are two types of operators, the linear operators constituted by the matrices A and R and the nonlinear operator $\theta\phi f$. The direction of flow in the diagram is the opposite of

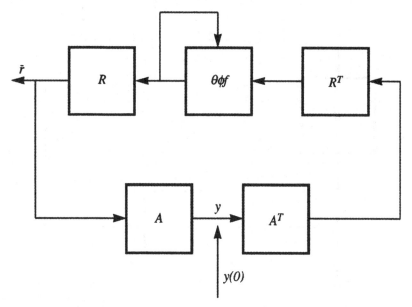

Fig. 8.3. A block representation of the static form (19), (20) of the probability-maximising algorithm (PMA). The side-loop to the $\theta\phi f$ block represents the determination of the standardisation factor θ by the rule (19), and the arrow marked $y(0)$ at the bottom represents injection of the initial value.

that usually taken in control contexts; we have taken this convention for consistency with those of the physiological block diagrams of Chapter 14. The side-loop associated with the ϕf block reflects the calculation (19) of the deflation factor θ from the output of this block. The side-arrow directed towards the link through which the signal y itself flows is intended to represent the mechanism by which the initial value of y is injected. See the Appendix for a caution on relation (8.20).

This next stage is represented in Fig. 8.4. We have rearranged the loop so that it begins to look more meaningful in a number of senses, including the physiological. The factors A and R have now been separated into the two blocks $A^T A$ and $R\phi f R^T$, which each have an embryonic H-net look about them. However, we have in fact taken the R factor so that it describes

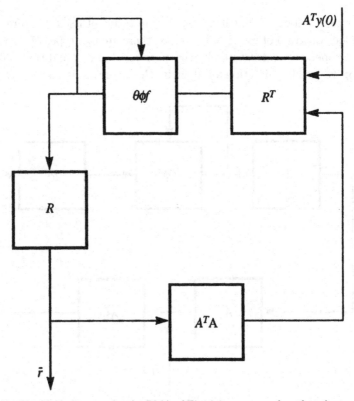

Fig. 8.4.   The block diagram for the PMA of Fig. 8.3, rearranged so that the two stages associated with the matrices A and R are separated. The output of the block R is continued as a system output. This carries the estimate $\bar{r}$ of the quantised weight vector r, and so constitutes the most meaningful reduction of the input data vector $y(0)$.

the multiple synaptic link running down the left-hand side of the diagram, and we have prolonged this link to indicate that it also carries the 'output' of the system, the estimate $\tilde{r}$. It is this estimate which provides the most meaningful p-dimensional characterisation of the data input $y(0)$.

Because input must now be injected on the other side of the A-block it is injected in the link carrying the signal $A^T y$, and so must be injected in the form $A^T y(0)$.

To complete specification we must return to the dynamic description, which leads us immediately to the consideration of neural implementation.

Note

The simplest extension of the simple-trace case is to admit the possibility of no trace at all: that $r = 0$. Suppose that this contingency has prior probability π_0, and that otherwise A, π and ϕ are defined as they were in relations (16) and (17). Then these equations are changed only in that (17) is modified to $\theta = (\pi_0 + 1^T \phi f A^T y)^{-1}$. The estimate of r is given by $\theta \phi f A^T y$.

8.6. Neural implementation of the PMA

We shall refer to the PMA in the simple-trace case $R = I$ as the *one-stage PMA* and in the case of general R as the *two-stage PMA*. This is because the factorisation AR in this second case really does invite separation of the net into two stages, as in Fig. 8.4.

A static scalar relation of the form $z = f(u)$ will be given the dynamic neural form

$$\mathcal{L}x = u, \quad z = f(x) \tag{8.21}$$

where $\mathcal{L}$ is the synaptic operator defined in Section 3.4, x is the membrane potential of the neuron and the activation function $f(x)$ gives the output pulse rate as a function of potential. (See Section 3.4 for a first discussion of these matters and Chapter 11 for a full treatment.) The activation function of actual neural masses and of plausible models for them has the familiar sigmoid form. We shall suppose that $f(x)$ represents the exponential $e^{x/v}$ of the last section well enough if it is operated well below saturation and can be correctly scaled.

The effect is then that a static relation $z = f(u)$ has to be seen in a dynamic and neural version as $z = f(\mathcal{L}^{-1}u)$, where by $\mathcal{L}^{-1}$ we mean the inverse in realisable form of the stable operator $\mathcal{L}$. (More strictly, it should be seen as $f(d^2 \mathcal{L}^{-1}u)$, where d^2 is the constant $\mathcal{L}1$, but for the moment we shall assume such constants absorbed in the definitions.) The dynamic and neural version of the block diagram of Fig. 8.4 then takes the form illustrated in Fig. 8.5. However, it is natural to add a few more features at this point.

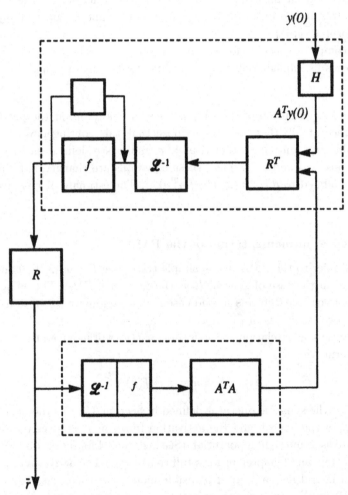

Fig. 8.5. The dynamic and neural realisation of the system of Fig. 8.4, defining the corresponding full realisation of the PMA. The sigmoid operator f now becomes $f\mathcal{L}^{-1}$, where $\mathcal{L}$ is the linear differential operator of equation (21). The side-loop to the top neural element again represents the feedback which effects standardisation, now realised by the determination (23) of an adaptive global bias q on the membrane potentials x. The injection of initial values is now represented more explicitly by the short-term holding unit H, which is also assumed to achieve the A^T transformation on the data. The top and bottom assemblies have been boxed in, to indicate their separateness.

If we are to interpret the system physiologically, as we shall, then it would be unnatural to imagine that the linear operation $R^T A^T A R$ performed in the return loop of the ϕf block is realised purely axonally. The path is a long one, with many transformations ('divergences' and 'convergences' in the axonal terminology) en route. We have therefore interposed a staging post: another neural element $f\mathcal{L}^{-1}$ in mid-loop.

Determination (19) of the deflation factor θ must also be given dynamic form. The elegant way to achieve this deflation, and also the ϕ-multiplication, is to feed the f-output back in such a way as to depress the neural potential x occurring in relation (21). Explicitly, suppose that we write the input and output of the vector neural element $\theta\phi f\mathcal{L}^{-1}$ as vectors u and z. We then incorporate the factors ϕ and θ by rewriting the relations as

$$\mathcal{L}x = u, \quad z_j = f(x_j - b_j - q). \tag{8.22}$$

Here x is the vector of potentials, b_j a fixed bias in potential which achieves the effect of multiplying f by ϕ_j and q a global adaptive bias which achieves the effect of multiplying f by θ. This alternative implementation is exact if f is truly exponential, but in most cases we shall be looking for the qualitative effect rather than the exact one. We suppose the corrective feedback relation for q

$$\dot{q} = \kappa'(\mathbf{1}^T z - 1) = \kappa'(\mathbf{1}^T f(x - b - q\mathbf{1}) - 1), \tag{8.23}$$

ensuring the normalisation (19) in equilibrium. Here κ' is an appropriate constant.

The side-loop to the $f\mathcal{L}^{-1}$ block in Fig. 8.5 represents the feedback relation (23). It is inserted between the two operations because it is indeed upon the intermediate variable x that it acts. This action is one of global inhibition, a feature observed physiologically and incorporated in one way or another into many models. In our case it follows necessarily from the deflation required by the PMA algorithm.

Next, we must consider how initial values (or changed data, indeed) are injected. In practice this cannot be achieved instantaneously. We have imagined that there is a 'holding unit' H which can retain incoming processed data $A^T y$ long enough to allow transfer of data into the net. This unit is in itself a memory, but one which both can and must be imperfect — it need only hold data until transfer has been effected, and it must clear itself to be able to receive new data in due course. It can be realised with internal-feedback units of the type specified in (3.13); we shall return to these matters in Chapter 14.

Finally, we have separately boxed in the top and bottom portions of the diagram, to label these as distinct sub-systems. Each is associated with something reminiscent of an H-net: the lower box concerned with recognising the presence of individual basic traces, the upper box concerned with identifying the weights with which these are combined. In Chapter 14, which attempts to relate models deduced to the anatomy of the olfactory system, we shall tentatively see these sub-systems as analogues of the anterior olfactory nucleus and the olfactory bulb. The physiological correspondence on a number of important features is quite striking.

The most explicit summary of the final structure of Fig. 8.5 is provided by the dynamic equations themselves. For the moment we shall assume that the holding unit H plays no part in these once the data have been injected, although that is an idea we shall revise in Chapter 14. Let x_1 and x_2 be the vectors of potentials for the upper and lower neural arrays respectively; these are of dimensions m and p respectively. The full set of dynamic relations is then

$$
\left.
\begin{aligned}
\mathcal{L}x_1 &= d^2 R^T W f(x_2) \\
\mathcal{L}x_2 &= d^2 R f(x_1 - b - q\mathbf{1}) \\
\mathcal{D}q &= \kappa[\mathbf{1}^T f(x_1 - b - q\mathbf{1}) - 1],
\end{aligned}
\right]
\tag{8.24}
$$

where $W = A^T A$ and we have now inserted the factor $d^2 = \mathcal{L}1$ to make relations correct in equilibrium.

8.7. The PMA and the exponential family

One may wonder to what degree the structure of the PMA (7) is specific to the Gaussian model (2) from which it was derived. We shall see that the structure survives for a wider class of models.

Suppose that the n-vector y has the probability density

$$
\psi(y \mid \alpha) = e^{\alpha^T y - P(\alpha) - Q(y)}
\tag{8.25}
$$

relative to Lebesgue measure on $\mathbb{R}^n$. This distribution is then a member of the exponential family, with variable y and parameter α so standardised that they appear in the combination $\alpha^T - y$. There will be such distributions which do not have a density relative to Lebesgue measure (e.g. the Poisson distribution); the results of this section extend to these if the notion of a derivative is appropriately modified.

A formal partial integration shows that

$$E[Q'(y)] = \alpha, \tag{8.26}$$

where Q' is the column vector $(\partial Q/\partial y_j)$.

Suppose now that the parameter α, an n-vector, can take the values a_k with prior probabilities π_k $(k = 1, 2, \ldots, p)$. We can regard these as the possible trace values, although the actual expectation of y if $\alpha = a_k$ is $P'(a_k)$ rather than a_k. The density of y, averaged over α, is then

$$\rho(y) = e^{-Q(y)} \sum_k \pi_k \exp[a_k^T y - P(a_k)].$$

We could write this as

$$\rho(y) = e^{-Q(y)}(1^T \phi f A^T y) = e^{-Q(y)}\theta^{-1},$$

say, where ϕ is the diagonal matrix with elements $\phi_k = \pi_k \exp[-P(a_k)]$, the function $f(x)$ is the vector of exponentials of the elements of the vector x, and A is the $n \times p$ matrix with columns a_k.

The equation derived from the probability-maximising principle would then be

$$\dot{y} = \kappa \frac{\partial \log \rho(y)}{\partial y} = \kappa[\theta A \phi f A^T y - Q'(y)].$$

This is almost identical with the Gaussian-derived PMA (7); the substantial differences are that $Q'(y)$ replaces y in the decay term and that, as we have stated, $E(y|\alpha)$ equals $P'(\alpha)$ rather than α. The equation could be written

$$\dot{y} = \kappa[E(\alpha|y) - Q'(y)].$$

The static version of this last equation presents an interesting (and well-known) dual to (26).

As in Section 5, one could go on to replace A by AR, so obtaining a compound-trace version of this non-Gaussian PMA.

8.8. Conclusions I

This is the best point at which to summarise the conclusions of Part II, since Chapter 9 is largely technical and Chapter 10 concerned with a balancing survey of related topics. Conclusions should be distinguished from opinions, but are sometimes less than novel in that they are largely a restatement and an emphasis of 'where we are'.

Recall the purpose of this Part, which was to find a rational basis for the design of an associative memory. Recall also our mixed attitude to the now classic Hopfield net: that it suffers from the various faults listed in Chapter 6, but nevertheless embodies something essential. Understanding of both its intended purpose and perceived faults has been incomplete.

We list the principal conclusions.

(1) An associative memory primarily performs an inference, but is converted by feedback to the secondary role of a storage memory. Storage in the presence of noise is possible only if the memory state is quantised.

(2) Incorporation of this second role is inevitable in the 'fading data' situation, when initial data cannot be stored, either inside or outside the network, without degradation. In this situation the operation of the memory is inevitably dynamic; inference cannot be performed instantaneously, and the dynamics must then be such that all important aspects of the data are retained as part of system state while the inference is being developed. It has to be accepted that the data are progressively lost: all that is left of them in the end is the inference itself, in the form of an idealised data vector.

(3) Most associative memories which have been proposed perform this dual role, although not designed with this explicit purpose in mind.

(4) The probability-maximising algorithm (PMA) provides a rationally-derived solution to the problem of inference with fading data. It also has asymptotically optimal properties, as we shall see in the next chapter.

(5) The PMA implies Hamming rather than Hopfield rules, amounting to a dynamic, autoassociative and somewhat blurred form of the Hamming net.

(6) The PMA also automatically incorporates a global inhibition in the form of a state-dependent scaling factor which largely suppresses all but the most active channel.

(7) The PMA separates the specification of basic traces and of the weights in which these can be combined. This leads to a two-stage operation in the feedback loop, which we shall later find highly reminiscent of some physiological structures.

(8) The PMA derivation makes plain that the design of the usual Hopfield net *implies* acceptance of the fact that composite traces can occur; it is then not surprising that the net shows a great number of 'spurious' equilibria. These can be eradicated, if one so wishes, only by designing on the basis of the traces, simple or composite, which one actually expects to occur.

(9) Operation with fading data inevitably implies the occurrence of positive feedback in the net, serving to reinforce the relevant features of the stored image. If the data are safely stored outside the net (and so available as a constant input) then feedback is negative rather than positive. In this case system state reflects discrepancies in explanation rather than the explanation itself (an idealised image of the data).

(10) Reliability considerations will force an upper bound on the basis rate $\alpha = p/n$ if the existence of p basic traces implies the existence of exponentially many (in p) compound traces. However, the consideration which sets this bound for the Hopfield net seems to be stability of recursive operation rather than reliability.

Chapter 9

Performance of the Probability-Maximising Algorithm

9.1. A general formulation

The approach which motivates the PMA seems so natural that one would conjecture its optimality in some sense, although ' optimality' is such a poorly defined concept in the neural context that one would be content to establish good performance. However, analysis is need to back up any such expectation, and it is only analysis which can reveal the operational factors.

We saw in Sections 8.5 and 8.6 that the PMA had an immediate and illuminating ANN realisation. It is moreover based upon a clear statistical extremal principle, which permits a relatively straightforward and illuminating statistical analysis. The replica method (for instance) is neither of these things and, for all its very considerable intrinsic interest, is probably not the natural tool in this context.

We shall take quite a general approach in this chapter, and shall consider the PMA based upon other models than the Gaussian model (8.2). We shall use the symbol ψ to denote the probability density (relative to an appropriate measure) of the random variable which it has as argument. So, $\psi(y)$ is the probability density of y, $\psi(y \mid x)$ is the probability density of y conditional on x, etc. The symbol ψ thus does not denote a specific function, but is simply read 'probability density of'.

The model specifies $\psi(y \mid s)$ in functional form, where y is the obervation vector and s the trace (simple or compound) which we believe lies behind that observation. For the reasons explained in Section 8.1, the associative memory must be based upon the assumption that s is quantised to a discrete set of m values, $s = Ar \, (r \in R)$ with respective prior probabilities π_r. The unconditioned probability density of y based upon this assumption is then

$$\rho(y) = \sum_r \pi_r \rho_r(y), \tag{9.1}$$

where $\rho_r(y) = \psi(y \mid r)$. We use the notation ρ rather than ψ, partly because we do now regard ρ and ρ_r as specific functions of their arguments, and partly because $\rho(y)$ is not the true density of y if s is not actually quantised. However, we *must* quantise in order to achieve stability of memory, and we suppose the r-values and the π_r chosen so that the quantised distribution approximates the actual one well. This is in itself an optimisation problem, addressed in the rate-distortion literature.

In studying performance we shall assume that the joint distribution of s and y implied in (1) *is* the true one, and that the performance of an inference procedure is judged on its performance for this quantised model. We shall then effectively assume this model the true one, and look for inference rules which are 'reliable', in the sense that r is correctly estimated from y with high probability, and 'optimal', in the sense that m is as large as it can be consistent with reliability. The inverted commas indicate incompleteness of definition, to be corrected in the next section.

We shall find (Section 3) that for simple stimuli and under appropriate hypotheses the PMA is asymptotically optimal, in that for large n it equals the performance of statistically optimal techniques. By the nature of the neural view this statement must be qualified: 'optimality' only under the assumption of a particular model, and an optimality which indeed implies the capability of distinguishing between exponentially many (in n) trace values, usually at the cost of an exponential amount of calculation.

In Sections 4 and 5 we study composite stimuli. As emphasised in Chapter 6, these are of interest, both because they presumably occur in nature and because they constitute the case in which full performance can be realised with economical calculation. With a basis of size p one has 2^p possible stimulus values, and so can hope to take p of order n, or possibly of higher order if the prior probability of a composite trace decreases rapidly enough with its degree of compositeness.

We do not allow for computation noise in the analysis, despite the fact that this was the feature which the PMA was intended to protect against. With properly chosen parameters the PMA protects against observation noise, in that the probability is high that the initial y lies in the correct basin of attraction of the algorithm. The expected time before y leaves this basin will be exponential in the reciprocal of computation noise power.

9.2. Considerations for reliable inference

As previously, we assume that there is a parameter n which measures the size of the data set. A given inference procedure (for the quantised model) is *reliable* if the probability of error in the identification of the trace tends to zero with increasing n.

We know from Note 2 of Section 2.1 that the procedure which minimises the probability of error is to choose the value of r which maximises $\pi_r \rho_r(y)$, with an arbitrary resolution of ties. So, suppose we consider the set A_r of y-values determined by

$$\pi_r \rho_r(y) > \max_{r' \neq r} \pi_{r'} \rho_{r'}(y). \tag{9.2}$$

This is an acceptance set for r, in that it is the set of y-values for which one would certainly decide for r. The sets A_r are disjoint, because we have excluded the cases of ties. For the same reason, they do not necessarily cover y-space, although they do so very nearly. The rule associated with (2) is optimal; it will also be reliable if $P(y \in A_r \,|\, r)$ tends to unity as $n \to \infty$, for all r in $\mathcal{R}$.

Consider now the rule implied by the PMA. Suppose that we can find a corresponding family of sets $\{\mathcal{B}_r\}$ with the properties that (i) the $\mathcal{B}_r$ are disjoint, (ii) $P(y \in B_r \,|\, r)$ tends to unity as $n \to \infty$ for all r in $\mathcal{R}$, and (iii) each B_r contains a single local maximum of $\rho(y)$, which is reached by the steepest ascent path of ρ starting from any point of B_r. If this can indeed be found, then the PMA is reliable, in that to each value of r corresponds a distinct value $\bar{y}_r$ of y (the maximising value in B_r) and that, if r is the true value, then the probability tends to unity with increasing n that the steepest ascent path from the data value y leads to $\bar{y}_r$. Convergence to $\bar{y}_r$ signals inference for the value r.

It is condition (iii) which raises issues entirely absent in application of the rule (2), in that it requires smoothness and unimodality of $\rho(y)$ within each B_r. However, we shall find that we can go a long way towards meeting these conditions if we specify B_r as the set of y for which

$$\pi_r \rho_r(y) > \delta^{-1} \sum_{r' \neq r} \pi_{r'} \rho_{r'}(y), \tag{9.3}$$

for a prescribed small constant δ. This is a very much more demanding condition than (2), a fact which implies incidentally that $\mathcal{B}_r \subset \mathcal{A}_r$ and that the $\mathcal{B}_r$ are disjoint. In fact, we shall find that the remaining reliability conditions on the $\mathcal{B}_r$ can to an extent be satisfied as easily as those on the $\mathcal{A}_r$, as must be the case if the PMA is to be asymptotically optimal.

The reasons for adopting the PMA rather than the Bayesian rule (2) are, of course, that it has a natural neural implementation, that it identifies the most compatible hypothesis by natural dynamics rather than by exhaustive enumeration, and that it resists degradation of the data during the inference calculations.

The calculation of the error probability is difficult for any inference rule, as we saw in Section 6.2, because the probability depends in a complicated fashion upon the specific traces s. We appealed already then to the communication-theory technique of 'random coding': that one chooses the traces at random from some distribution, and calculates the probability of error under variation of these traces as well as under variation of observation error and of the trace which is actually presented on a given trial. This has a symmetrising effect, and expressions are much more amenable. It may also make more sense, in that one considers average performance over a class of cases rather than for one particular case. (Note that trace variation then has two levels, which should be distinguished. One is the possible random choice of the m traces which are to represent the m stimuli; once made, this is fixed. The other is the variation of trace presented from trial to trial, as the stimulus also varies.)

The analysis of Section 6.2 leads us to expect that m should be able to grow exponentially fast with n consistent with reliability; as $\exp(Kn)$, say. Denote the supremum value of such K by C; this is then a *capacity*, in nats of information per element of y. The value of the capacity will depend upon the inference rule used; let us denote the capacity for the optimal rule (2) by C, and for the probability-maximising algorithm by C_{PM}. Since the Bayesian rule is optimal we have

$$C_{PM} \leq C. \qquad (9.4)$$

The capacity will also depend upon the conditional density $\psi(y \mid s)$ and the way in which it varies with n, the prior probabilities π_r, and the way they vary with m, and the generating distribution $\psi(s_1, s_2, \ldots, s_m)$ by which the stimuli s_j are chosen. In the communication context the random coding is actually chosen to optimise capacity, since one then has the freedom to choose the transmitted signals (traces) which are to represent messages (stimuli). In the neural context one does not have that latitude. Nature determines stimulus statistics, although rather artificial statistics are sometimes assumed in order to permit evaluation of performance. Moreover, as we have repeatedly emphasised, although reliability considerations may permit exponentially many trace values, computing considerations will in general set a much more drastic limit unless these values show a particular structure.

9.3. Performance of the PMA for simple stimuli

The stimuli s may be compound, but let us first consider the case where they are simple, so that $s_j = a_j$ $(j = 1, 2, \ldots, m)$ where a_j is a simple memory trace. The 'simplicity' lies in the fact that the traces are unrelated to each other, functionally or statistically. Indeed, we shall make the assumption, perhaps not physically realistic, but familiar in analyses of performance, that the m traces a_j have been chosen as independent realisations of a random vector a with prescribed distribution.

If we know that the stimulus could take one of the m values a_j with respective prior probabilities π_j then the inference problem is that of determining which of these stimuli is most consistent with the data vector y. As indicated in Chapter 6 and in the last section, this is not only the classical hypothesis-testing problem, it is also the classical statistical problem of information transmission through a noisy channel (see e.g. Blahut (1987) or Goldie and Pinch (1991)). A message can take one of m values, the jth is represented by transmission of the signal a_j, its n elements constituting a sequence transmitted over n units of time. The data vector y is just the signal received at the other end of the channel, and the channel is characterised statistically by the form of the probability density $\psi(y \mid a)$ of output y conditional on input a. This is plainly just a rephrasing of the inference problem: the jth hypothesis is that the 'word' a_j was transmitted $(j = 1, 2, \ldots, m)$.

The communication interpretation raises the same consideration of reliability as does the hypothesis-testing interpretation; it also raises the considerations of random coding and capacity which we touched on in the last section. If m is allowed to increase with n then the set of traces a_j and their corresponding prior probabilities π_j must also be varying. One usually achieves this in performance studies by choosing the a_j as m independent draws from a common a-distribution, and by assuming the stimuli equiprobable: $\pi_j = 1/m$. This latter choice is the simplest, and is also conservative, in that it represents a worst case as far as error probability is concerned.

As in the last section, we shall denote the capacities achieved by the optimal rule (2) and the probability-maximising algorithm by C and C_{PM}

respectively. We now consider these for the 'classical' case of simple traces, generated as independently and identically distributed random vectors in the random coding.

We shall replace notations such as ρ_r and $\mathcal{B}_r$ by ρ_j and $\mathcal{B}_j$, although consistency would demand that we replaced r by e_j, where e_j is the vector with a unit in the jth place and zeros elsewhere. We shall regard y and a as having a joint probability density $\psi(y \mid a)\psi(a)$, specified by the channel rule and the coding rule, and $\psi(y)$ as being the corresponding marginal density of y.

Theorem 9.1. *Define the random variable* $\zeta = n^{-1} \log[\psi(y \mid a)/\psi(y)]$. *Suppose the m traces equiprobable and that ζ converges in probability to a constant γ with increasing n. Then*

(i) *The PMA is reliable to the extent that, for the sets $\mathcal{B}_j$ defined by (3), the probability $P(y \in \mathcal{B}_j \mid s = a_j)$ tends to unity with increasing n and*

$$\rho_j(y) \leq \rho(y) \leq (1+\delta)\rho_j(y) \quad (y \in \mathcal{B}_j). \tag{9.5}$$

(ii) *The PMA is optimal to the extent that $C = \gamma$ and that assertion (i) continues to hold if $\log m$ grows at a rate $(\gamma - \eta)n$ for any prescribed positive η.*

The reliability and capacity properties thus certainly hold for the acceptance sets $\mathcal{B}_j$. The reason for the degree of reservation in the statement of the theorem is that the third requirement placed on these sets in the discussion of the last section (that each $\mathcal{B}_j$ should contain a single local maximum of $\rho(y)$, which is reached by the steepest ascent path of ρ starting from any point of $\mathcal{B}_j$) is satisfied only to the extent that (5) holds, for arbitrarily small but fixed δ. Relation (5) would imply these conclusions if $\rho_j(y)$ were itself unimodal and $\rho(y)$ were sufficiently smooth in $\mathcal{B}_j$, conditions which we would expect to hold in most cases of interest. However, we shall not continue the analysis of this point.

The formula for ζ defines this variable as a specific function of a and y and so is itself a random variable. Readers familiar with statistical communication theory will recognise the Shannon evaluation of capacity in the above evaluation of C (apart from the fact that coding is not optimised). The point of the theorem is that the PMA also realises this asymptotically optimal performance, at least if subsidiary unimodality/smoothness conditions are satisfied.

If the stochastic convergence required in the theorem holds, then under mild conditions one can make the identification

$$\gamma = \lim_{n \to \infty} E\zeta = \lim_{n \to \infty} n^{-1} i(a, y). \tag{9.6}$$

where $i(a, y)$ is the mutual information between a and y, calculated under the joint probability density $\psi(a)\psi(y \,|\, a)$.

Proof of Theorem 9.1. Relation (5) is an immediate consequence of the defining condition (3); it is fulfilment of this condition with a probability converging to unity which must be established. Let us write, as before, $\rho_j(y) = \psi(y \,|\, a = a_j)$ and define

$$\xi(y) = \sum_{k \neq j} \rho_k(y). \tag{9.7}$$

The expectation of $\xi(y)$ over variation of the traces a_k $(k \neq j)$ is then just $(m-1)\psi(y)$. We can establish reliability if we can demonstrate that $P(\mathcal{E})$ converges to zero with increasing n, where $\mathcal{E}$ is the event $\psi(y \,|\, a) < \delta^{-1}\xi(y)$. Here we have written a_j just as a.

Appealing to Markov's inequality we have

$$P(\mathcal{E} \,|\, a, y) \leq \min \left[\frac{E[\xi(y) \,|\, a, y]}{\delta\psi(y \,|\, a)}, 1 \right]$$

$$\leq \min \left[\frac{m\psi(y)}{\delta\psi(y \,|\, a)}, 1 \right]$$

$$= \min[m\delta^{-1}e^{-n\zeta}, 1] \tag{9.8}$$

It follows from (8) that, for any prescribed positive η,

$$P(\mathcal{E}) < me^{-n(\gamma-\eta)-\log\delta} + P(\zeta < \gamma - \eta).$$

The final probability converges to zero with increasing n, by hypothesis, and so the whole expression will converge to zero if $m \sim \exp(Kn)$ and $K < \gamma-\eta$. Since both η and the margin of inequality can be chosen arbitrarily small, this establishes γ as a lower bound on C_{PM}. The reverse inequality (4) then implies equality. $\qquad \square$

The asymptotic inequality

$$P(\mathcal{E}) < me^{-n\gamma} = me^{-i(a,y)}$$

constitutes a form of Stein's Lemma (see Blahut, 1987), with the additional feature of an averaging over trace variation.

It is evident from the proof of the theorem that the reliability and capacity conclusions would still hold if δ decreased with n at a rate slower than exponential. It may seem strange that these sharp conclusions can be reached by use of the crude Markov inequality. The reason seems to be that, for values of m consistent with reliability, the traces a_j are too sparsely distributed in a-space that $\xi(y)$ (for given y) can be normally distributed in the limit of large n. The distribution of $\xi(y)$ is then such that the Markov inequality is not so crude.

One case of interest is then the simple additive model $y = a + \epsilon$ of Chapter 5, with the elements of the noise vector ϵ supposed independently and normally distributed with zero mean and variance v. Suppose the traces generated by choosing the mn elements a_{jk} of the m traces as independent realisations of a normally distributed random variable with zero mean and variance P. Then we find readily that (6) reduces to

$$C_{PM} = C = \frac{1}{2} \log \left[1 + \frac{P}{v} \right].\tag{9.9}$$

This is the classic evaluation of the capacity of a memoryless Gaussian channel, because the coding suggested is in fact optimal under the constraint $E(a_{jk}^2) < P$ (see the references quoted).

Another standard case is that for which all vectors are binary, in that their elements take the values ± 1, say, and it is supposed that errors in the different elements of y are independent and have probability μ. Suppose also the random coding in which the a_{jk} are independent realisations of a random variable taking the values ± 1 with equal probability. One has then the evaluation

$$C_{PM} = C = \log 2 + \mu \log \mu + (1 - \mu) \log(l - \mu).$$

This is the standard capacity evaluation for the 'binary symmetric channel', since the coding suggested is again optimal.

9.4. Compound stimuli: The general pattern

We return to the additive compounding of p basic traces, so that a trace s takes the form Ar, where r is the p-vector of loadings. The probability density of the data vector y conditional on the value s of the stimulus transmitted is then $\psi(y \mid s)$.

The value of r must be quantised; we shall consider choices later in the section. Let s and s' denote compound traces corresponding to loadings r and r' respectively. The argument adopted for the single-trace case in the last section generalises. We wish to determine an upper bound on the probability of the event $\mathcal{E}$ that

$$\pi_r \psi(y \mid s) < \delta^{-1} \sum_{r' \neq r} \pi_{r'} \psi(y \mid s') \qquad (9.10)$$

where s is the true trace. Define

$$\psi_s(y \mid s') = E_{s'}[\psi(y \mid s') \mid s],$$

where the expectation is over the *coding* of the traces in s'. The conditioning on the value of s will then imply that the only traces whose coding is averaged are those which do not also appear in s. Then it follows as in (8) and the inequality following it that

$$P(\mathcal{E} \mid s, y) \leq \delta^{-1} \sum_{r' \neq r} \frac{\pi_{r'} \psi_s(y \mid s')}{\pi_r \psi(y \mid s)} \approx \delta^{-1} \sum_{r' \neq r} (\pi_{r'}/\pi_r) C^{-nC(r,r')}, \quad (9.11)$$

where we have assumed that $n^{-1} \log[\psi(y \mid s)/\psi_s(y \mid s')]$, a function of s and y, has a stochastic limit $C(r, r')$ under coding variation of s and sampling variation of y. Averaging over r, we deduce inequality (12) below, expressing a conclusion which we state formally.

Theorem 9.2. *If the quantity $n^{-1} \log[\psi(y \mid s)\psi_s(y \mid s')]$ indeed has a stochastic limit $C(r, r')$ under coding variation of s and sampling variation of y then the bound*

$$P(\mathcal{E}) \leq \delta^{-1} \sum_{r' \neq r} \sum \pi_{r'} e^{-nC(r,r')} \qquad (9.12)$$

holds for the probability $P(\mathcal{E})$ that the PMA leads to a false conclusion. More precisely, the probability that y lies in the correct $\mathcal{B}_r$ and that

$$\rho_r(y) \leq \rho(y) \leq (1 + \delta)\rho_r(y)$$

within this set exceeds $1 - P'$, where P' is the bound asserted in (12).

We should now consider specific cases. One case of interest for illustrative purposes, at least, is that in which the prior distribution π_r specifies the elements r_i of r as being independently and identically distributed random variables, taking the values 0 and 1 with probabilities

λ and μ. Thus $\pi_r = \lambda^{p-j}\mu^j$, if j is the number of active traces (i.e. the number of simple traces present in s, which we shall also term the size of s).

Suppose that s and s' are of sizes j and k respectively, and have h traces in common. Then $C(r, r')$ is a function C_{hjk} of these variables alone. In terms of these variables the sum (12) will undergo considerable condensation. There are $\binom{p}{j}$ traces of size j and $\binom{p-j}{k-h}\binom{j}{h}$ traces of size k which have an overlap of size h with a given trace of the first group; the product of these expressions gives the multiplicity of the hjk term. We find then that the bound (12) becomes

$$P(\mathcal{E}) \le \delta^{-1} \sum_h \sum_j \sum_k \frac{p!(\lambda^{p-k}\nu^k)}{h!(j-h)!(k-h)!(p-j-k+h)!} e^{-nC_{hjk}},$$

(9.13)

where, as well as the obvious restrictions on h, j and k (i.e. that all the factorial expressions have non-negative argument), combinations with $h = j = k$ are excluded from the summation.

Before we begin to consider more explicit cases, there is a further point to be made. We expect p to be able to grow with n, at a rate yet to be determined. However, the assumption that compounding can be continued to any degree is rarely one that can then be sustained. Indefinite compounding with increasing p leads to increasing magnitude of the elements of s. This has the effect that either the elements of y must grow correspondingly (as would be the case for the linear model $y = Ar + \epsilon$) or, if the elements of y are bounded, then the elements of y will be pressed against their upper bound with increasing probability. That is, there will either be an uncontrolled growth in signal amplitude or saturation. Either contingency is unacceptable.

Realism is restored if we suppose μ to be of order p^{-1}, so that $\mu = \nu/p$, say, for some constant ν. In the limit of large p the number of active traces is then Poisson distributed with expectation v.

With this assumption, and the assumption of large p, relation (13) becomes

$$P(\mathcal{E}) \le \delta^{-1} \sum_h \sum_j \sum_k \frac{p^{j-h}e^{-\nu}\nu^k}{h!(j-h)!(k-h)!} e^{-nC_{hjk}}, \qquad (9.14)$$

with the same restrictions on h, j and k as previously. Note one point of interest. One expects that C_{hjk} will increase as h decreases, for given j

and k, because, the less the overlap between the two traces s and s', the easier it is to discriminate between them. The exponential term in (13) thus decreases faster with n as h decreases. However, the term p^{j-h} increases — a combinatorial effect. A decrease in prescribed overlap makes discrimination easier (between a given pair of traces of sizes j and k) but greatly increases the number of admissible k-traces.

The hjk term in this sum will tend to zero with increasing n if

$$(j - h) \log p \leq n(C_{hjk} - \eta)$$

for a fixed positive η. We can thus deduce:

Theorem 9.3. *The PMA is reliable for the model with Poisson compounding if p grows with n as e^{Kn}, where*

$$K < \inf_{h,j,k} C_{hjk}/(j - h). \tag{9.15}$$

Here j and k (and consequently h) are restricted to a set $j, k \leq M$ with h, j and k not all equal, and M is chosen so that the prior probability that a compound trace contains more than M basic traces is negligible.

This effective upper bound on the values of h, j and k implies that the bound (14) is effectively polynomial in p, and it is this which allows p to grow exponentially fast with n, consistent with reliability. If we return to the case (13) of unrestricted compounding then j and k cannot be bounded in this way; the permitted rate of growth of p is then greatly decreased.

One particular such case which we shall consider in the next section is that in which the true trace is simple, although compounding is unrestricted. The event of error in this case amounts to the event that a single trace is wrongly inferred when a full array of 'spurious' equilibria exist. This corresponds to the case $j = 1$, and the hjk reduction of the bound (11) then becomes

$$P(\mathcal{E}) < \delta^{-1} \sum_{k=0}^{p-1} \binom{p-1}{k} (\mu/\lambda)^{k-1} e^{-nC_{01k}}$$

$$+ \delta^{-1} \sum_{k=2}^{p} \binom{p-1}{k-1} (\mu/\lambda)^{k-1} e^{-nC_{11k}} \tag{9.16}$$

where the two sums correspond to $h = 0$ and $h = 1$ respectively.

9.5. Compound stimuli in the Gaussian case

The choice of a stochastic model is equivalent to specification of $\psi(y \mid s)$; this then in principle determines the exponent C_{hjk} of (13) and (14). The aim is to determine the greatest rate of growth of p with n which will ensure that the dominant term in these sums tends to zero with increasing n.

We assume the Gaussian model of Section 3, but with compound traces, so that $y = s + \epsilon = Ar + \epsilon$. Then

$$\psi(y \mid s) = (2\pi v)^{-n/2} \exp[-|y - s|^2/2v]$$

$$\psi(s) = (2\pi v)^{-nj/2} \exp\left[-\sum_i |a_i|^2/2P\right],$$

where the sum over i covers the j traces a_i which occur in s.

One finds then that

$$\psi_s(y \mid s') = [2\pi(v + (k - h)P)]^{-n/2} \exp\left[-\frac{|y - s''|^2}{2(v + (k - h)P)}\right],$$

where s'' is the common part of traces s and s', composed of h simple traces. If we set $y = s + \epsilon$ in $n^{-1} \log[\psi(y \mid s)/\psi_s(y|s')]$ then we see that this expression has the stochastic limit

$$C_{hjk} = \frac{1}{2} \log[1 + (k - h)\Gamma] + \frac{(j - k)\Gamma}{2[1 + (k - h)\Gamma]}, \tag{9.17}$$

where Γ is the signal-to-noise ratio P/v. Expression (17) has its minimal value of zero in the forbidden case $h = j = k$.

The bound (15) on the permitted exponent K in $p = e^{Kn}$ is then a function only of $j - h$ and $k - h$, so we may as well set $h = 0$. Expression (17) is then minimal with respect to k at $k = j$; the bound (15) on K then becomes $\inf_j (2j)^{-1} \log(1 + j\Gamma)$. We thus deduce;

Theorem 9.4. *Consider the Gaussian model with Poisson compounding, a signal/noise ratio of P/v and an effective bound M on degree of compounding. The PMA is reliable for this model when p grows with n as e^{Kn}, for $K < (2M)^{-1} \log(1 + MP/v)$. It thus achieves a capacity of*

$$C = \frac{1}{2} \log(1 + MP/v).$$

This indeed reduces in the case $M = 1$ to the classic evaluation (9) for the single-trace case.

Proof. The number of compound traces which can be reliably distinguished is $\sum_{k=0}^{M} \binom{p}{k} = O(p^M) = e^{MKn+o(n)}$, and the expression given for C is just our evaluation of the supremal value of MK. □

The guillotining of the size distribution at M amounts to saying that one excludes errors which might be made if size j exceeds M. However, if one wishes the probability of any kind of error (at a rate which has been bounded for $j \leq M$ and not bounded for $j > M$) to tend to zero with increasing n then one must let M tend to infinity with n, although it can do so at an arbitrarily small rate. One can say, then, that reliability can be achieved if $\log p$ increases with n as $[n/2M(n)] \log(l + M(n)P/v)$, where $M(n)$ is a function which increases indefinitely with increasing n. That is, reliability will be achieved if $\log p$ is $o(n)$ for large n, so that $\log p$ grows with n at any sublinear rate. A more careful analysis of the terms in the sum (14) confirms this.

Finally, let us attempt to evaluate bound (16). As mentioned, this is a bound for the probability that a single trace is wrongly identified against a full array of indefinitely compounded alternatives, and so of spurious traces. The kth term in both sums of (16) has the order of magnitude of

$$Q_k := \binom{p}{k}(\lambda/\mu)^k(1 + k\Gamma)^{-n/2}e^{n/2}$$

if, as we may expect, the relevant k is large enough that the contribution from the final term in expression (17) for C_{hjk} can be taken as $-n/2$. We require to know the largest rate of growth of p with n such that Q_k will tend to zero with increasing n for all $k \leq p$.

If we assume k and p to be both of order n then we find that Q_k is of order $n^{-n/2}$, so that this is certainly not the unfavourable case we are committed to finding. If we assume k to be of order p, and p to be of order greater than n, then we find that Q_k is maximal for $k = \lambda p$, and is then asymptotic to

$$\mu^{-p}(1 + \lambda\Gamma p)^{-n/2}e^{n/2} \sim e^{c_1 p/2}(c_2 p)^{-n/2}.$$

where $c_1 = -2\log\mu$ and $c_2 = \lambda/e$. The transitional case, when this is $O(1)$, corresponds to $n = c_1 p/\log(c_2 p)$, or

$$p = \frac{n \log n}{2 \log \mu} + o(n \log n).$$

Reliability thus demands that p should grow no faster than $n \log n$.

Chapter 10

Other Memories — Other Considerations

The reader may regard this chapter as an optional excursion and rest-point.

The aim of the text is to find a rational basis for the construction of an associative memory and, for reasons and by means described in the next part, to adapt this to include the additional feature of oscillatory operation. However, in following this path over-intently we neglect a great deal of interesting, fundamental and relevant material. In particular, we neglect the features of learning (perhaps the prime motivation of ANN development) and of techniques derived, not from optimisation of a linear model, but from very general concepts of reduction and recognition. In this chapter we attempt to restore balance by quickly surveying some of this material before resuming the quest.

References are given at various points, but a text such as that by Hassoun (1995) covers much of the material.

10.1. The supervised learning of a linear relation

We have clung strongly to the simple linear model $y = Ar + \epsilon$, relating the data of a stimulus y to the reality r of whatever caused the stimulus. The assumption has been throughout that the matrix A was known, and one may ask how this knowledge was acquired. In *supervised learning* one is presented with a sequence of cases in which the values of both y and r are given, so that one has a basis from which to infer the value of A.

This is by no means an artificial situation. In real life the cause of a stimulus is often evident at more or less the same time as its effects, as we argued in Chapter 7. For example, an animal coming across a scent may soon catch up with its cause, and learn to associate the two. In picking up the scent it has observed y; by seeing that it was caused by a flock of sheep accompanied by a dog and a shepherd it has observed r. Supervised learning is thus provided in this case by the supplementation of the olfactory information by visual information; the learning is the learning of the association.

Actually, the animal (even if unnaturally rational) will not go through the steps of inferring A (from the sequence of supervised trials) and then solving the equation system (6.10) for the estimate of r in the next unsupervised trial. Rather, it will learn from the supervised trials to construct a direct estimate $\hat{r} = By$. By an optimal determination of B it has effectively inverted the matrix $A^T A$ (or a Bayesian equivalent) once and for all.

One can easily develop a simple rule for real-time revision of the estimate of B which illustrates at least one learning principle. This is the principle of gradient descent, an empirical version of steepest descent minimisation of a cost function. The approach may sound rather premeditated, but the learning rule derived often seems to be very natural and to stand on its own merits.

In the present case the aim is to choose a matrix B which minimises the loss function $C(B) = E[|v - By|^2]$, say. The stationarity condition is of course a simple linear equation in B with various covariance matrices as coefficients. However, one wishes to find a simple way of converging on to the solution of this equation without explicitly storing or inverting these covariance matrices.

In a discrete version of steepest descent minimisation one would change B by a sequence of increments

$$\delta B = -\kappa \frac{\partial C}{\partial B} = \kappa E[(r - By)y^T]. \qquad (10.1)$$

where κ is a suitably small constant. An empirical equivalent of (1) is

$$\delta B = \kappa(r - By)y^T, \qquad (10.2)$$

where δB is the change made to B after a new trial, at which the values y and r are observed.

Relation (2) constitutes the learning rule. It has relations to the so-called Hebbian learning rule: see Section 5. It requires storage only of the current value of B, whose elements are stored as the synaptic strengths of the net which will serve to calculate the estimate $\hat{r} = By$ when supervised learning ceases. The constant κ must be chosen small enough that the coefficients ry^T and yy^T occurring in (2) have time to average out statistically before B changes too much.

In time B will converge to the value $V_{ry}V_{yy}^{-1}$, which has the value $UA^T(AUA)^T + V)^{-1}$ if v and ϵ have covariance matrices U and V respectively, and are mutually uncorrelated. This in fact agrees with the evaluation $(U^{-1} + A^TV^{-1}A)^{-1}A^TV^{-1}$ derived by the maximisation of the joint density of r and y with respect to r; the two forms amount to equivalent reductions of an equation system. But, of course, the algorithm (2) does nothing so explicit or brutal; it just converges to the correct value, storing past data only in the current evaluation of B.

A network realising the linear rule $\hat{r} = By$ appears somewhat skimpy as a neural net, in that it lacks the nonlinear activation functions so characteristic of a neuron. This is because the rule is indeed linear — a sigmoid activation function is necessary (and also sufficient lor approximate realisation) if one wishes to realise nonlinear rules. However, the learning rule (2) for the synaptic weights of the net is nontrivial, and exemplifies more general cases.

10.2. Unsupervised learning: The criterion of economy

Keeping still with the linear model, let us suppose that supervised learning is not available, so that no information is available other than the sequence of y observations. One has then no other course than to look for statistical regularities of the random vector y. The model $y = Ar + \epsilon$ implies such a regularity, in that it implies that y is confined, to within the fuzziness of observation error, to a linear manifold of dimension p. Can one infer such an approximate dimensional reduction from repeated trials?

The ANN view of such a reduction is as follows. Suppose one reduces y to a p-dimensional vector x by the transformation $x = By$. Suppose one then tries to reconstruct an estimate $\hat{y}$ of y from x by the transformation

$$\hat{y} = Ax = ABy. \qquad (10.3)$$

The natural question is: how should one choose the $n \times p$ and $p \times n$ matrices A and B to achieve the best possible reconstruction for given p? The solution is plainly indeterminate, since one can as well choose AJ and $J^{-1}B$ as A and B, for a nonsingular $p \times p$ matrix J.

However, the criteria are clear: they are those of *economy* (in that one would like p to be substantially smaller than n) and *faithfulness* (in that $\hat{y}$

should approximate y well). We should now consider what performance is possible before considering how it could be realised by a learned reduction.

The variable x plays something of the role of the r of the linear model: the r-vector whose value essentially determines the value of the n-vector y. However, we now have no r to observe, so must essentially generate it from y itself, in the form of $x = By$.

This formulation reflects a mechanism which we have already seen in the associative memories of Section 5.3, in which a memory is alternately stored in two representations: a condensed 'abstract' representation and an expanded 'sensory' representation. This seems to be a natural mode of operation, the abstract phase often being associated with a digestion of impressions received and the sensory phase with a checking of recall against reality. It recurs constantly in ANN theory; e.g. in the Helmholtz machine and in the Grossberg adaptive resonance mechanisms of Section 7.

10.3. Principal components

Suppose that all we have learned from long observation is that the data vector y has covariance matrix V. (For simplicity, we shall suppose that all variables have been reduced to zero mean.) Suppose we consider a scalar linear form $\xi = c^T y$ whose coefficient vector c is normalised to unit length (i.e. $c^T c = 1$), and look for the form whose mean-square value $E(\xi^2) = c^T V c$ is maximal or minimal. If we look for the value which is maximal then we determine the 'component' of y which shows greatest variability. If we look for the value which is minimal then we determine the 'component' of y which shows least variability; i.e. the hyperplane in $\mathbb{R}^n$ to which y is most closely confined.

The stationarity condition with respect to c is $Vc = \lambda c$, with $\lambda = E(\xi^2)$. That is, c is a normalised eigenvector of V, with $E(\xi^2)$ the corresponding eigenvalue. One thus arrives at n such forms, which we shall denote by $\xi_j (j = 1, 2, \ldots, n)$ and which are termed the *principal components* of y. Note that the eigenvalues λ are real and non-negative and that the eigenvectors c can be taken as real and mutually orthogonal.

The following theorem presents a stronger statement, whose proof we leave to the reader.

Theorem 10.1. *The principal components are defined as the linear forms* $\xi_j = c_j^T y$, *where c_j is a normalised eigenvector of V, with corresponding* *eigenvalue* λ_j. *These satisfy the statistical orthogonality relationship*

$$E(\xi_j \xi_k) = \lambda_j \delta_{jk}. \tag{10.4}$$

If the eigenvalues are ordered $\lambda_1 \geq \lambda_2 \geq \cdots \geq \lambda_n$ *then we can characterise* ξ_j *as the normalised linear form whose variance is maximal subject to the constraint that* ξ_j *is statistically orthogonal to* $\xi_k (k < j)$.

Note the representations

$$V = \sum_{j=1}^{n} \lambda_j c_j c_j^T, \quad y = \sum_{j=1}^{n} \xi_j c_j, \tag{10.5}$$

the so-called *spectral representations* of covariance and variable. The reduction/reconstruction problem of (3) has an immediate solution in terms of these.

Theorem 10.2. *The p-dimensional representation* (3) *which minimises* $E[|\hat{y} - y|^2]$ *is* $\hat{y} = y_{(p)}$ *where* $y_{(p)}$ *is the curtailed version*

$$y_{(p)} = \sum_{j=1}^{p} \xi_j c_j \tag{10.6}$$

of the spectral representation (5).

It is convenient to prove first:

Lemma 10.3. *Let D he an* $n \times p$ *matrix. Then*

$$\mathrm{tr}[(D^T D)^{-1}(D^T V D)] \leq \sum_{j=1}^{p} \lambda_j, \tag{10.7}$$

with equality when D has columns $c_j (j = 1, 2, \ldots, p)$.

Proof of the Lemma. Write the trace expression as $\mathrm{tr}[Q^{-1}P]$. We can bring P and Q simultaneously to diagonal form by a nonsingular transformation $D \to DJ$ without changing the value of this expression. If we then extremise the trace expression with respect to D we derive the stationarity condition $VD = D\Lambda_p$, where Λ_p is the diagonal matrix PQ^{-1}. This implies that the columns of D are eigenvector of V, necessarily distinct if Q is to be nonsingular, and that Λ_p is the diagonal matrix of p corresponding eigenvalues. Further, $\mathrm{tr}(Q^{-1}P) = \mathrm{tr}(\Lambda_p)$ is the sum of these eigenvalues. Inequality (7) and the case of equality thus follow. □

Proof of Theorem 10.2. For given B one verifies that the optimal value of A is $V_{yx}V_{xx}^{-1} = VB^T(BVB^T)^{-1}$ and we have

$$
\begin{aligned}
E[|\hat{y} - y|^2] &= \operatorname{tr}[V_{yy} - V_{yx}V_{xx}^{-1}V_{xy}] \\
&= \operatorname{tr}[V - VB^T(BVB^T)^{-1}BV] \\
&= \operatorname{tr}[V - P_1^{-1}P_2],
\end{aligned}
\tag{10.8}
$$

where $P_s = BV^S B^T$. By Lemma 10.3 this implies that

$$
E[|\hat{y} - y|^2] \geq \sum_{j>p} \lambda_j.
$$

But this bound is attained by $\hat{y} = y_{(p)}$.   □

The curtailed representation $y_{(p)}$ is indeed a p-dimensional approximation to y. It has the symmetric form $AA^T y$, where A is the matrix whose columns are the first p eigenvalues of V.

The following lemma, an advance on Lemma 10.3, will prove useful in the next section.

Lemma 10.4. *Let D be an $n \times p$ matrix. Then*

$$
G(D) := \operatorname{tr}\{(D^T D)^{-1}(D^T V^2 D) - [(D^T D)^{-1}(D^T VD)]^2\} \geq 0. \tag{10.9}
$$

If the eigenvalues of V are distinct then equality can hold in (9) if and only if D is of the form $H_p J$, where the columns of H_p are p distinct eigenvectors of V and the $p \times p$ matrix J is nonsingular.

Proof. Consider first the case $p = 1$, when D is an n-vector d, say. Appealing to the spectral representation $V^s = \sum_j \lambda_j^s c_j c_j^T$ we have then

$$
G(D) = G(d) = \sum_j \pi_j \lambda_j^2 - \left(\sum_j \pi_j \lambda_j\right)^2, \tag{10.10}
$$

where $\pi_j = (d^T c_j)^2 / |d|^2$. The π_j thus constitute a distribution, and expression (10) can be zero if and only this distribution is concentrated on a single j value; i.e. if d is proportional to one of the eigenvectors c_j.

For $p > 1$, we appeal to the normalisation of D in Lemma 10.3 which simultaneously diagonalised $D^T D$ and $D^T VD$. If the normalised D has columns d_j then one finds that $G(D) = \sum_j G(d_j) \geq 0$. Equality can hold if and only if every d_j is proportional to an eigenvector of V, and these must be distinct eigenvectors if D is to be nonsingular.   □

10.4. The learning of an optimal reduction

Remarkably, there is a simple learning rule which generates the optimal reduction $x = A^T y$ and so effectively identifies the space spanned by the first p principal components. This was demonstrated first for $p = 1$ by Oja (1982) and then for general p by a sequence of authors (Karhunen and Oja (1982), Sanger (1989), Baldi (1989), Oja (1989, 1992), Hornik and Kuan (1992)).

The motivation for the rule is not quite immediate; we give it at the end of the section. The rule is

$$\delta A = \kappa(y - Ax)x^T, \quad \text{where } x = A^T y. \tag{10.11}$$

We could have eliminated x from these relations to express the increment δA purely in terms of y, but the roles of the various terms are clearer if x is retained, as is network realisation.

If κ is small enough that averaging takes place before A changes substantially then a locally time-averaged version of equation (11) is

$$\delta A = \kappa(VA - AP_1), \tag{10.12}$$

where $P_s = A^T V^s A$. Relation (12) is essentially the deterministic version of the stochastic relation (11). The replacement of one by the other would need a more careful discussion, but it is indeed clear that the approximation is justified for small κ.

Theorem 10.5. *The learning rule* (12) *leads from almost all initial values of A to a reduced form x which is a linear invertible function of the first p principal components of y, and so to the reconstruction $\hat{y} = y_{(p)}$.*

That is, it leads to a p-dimensional reconstruction of y which we know to be optimal: that yielded by the curtailed principal-component expansion (6).

Proof. It follows from (12) that

$$\dot{P}_s = 2P_{s+1} - P_1 P_s - P_s P_l. \tag{10.13}$$

Now, if the reduction $x = A^T y$ is adopted, then it follows from equation (8) that we can achieve a reconstruction with a mean-square error of $\text{tr}(V) - R$, where $R = \text{tr}(P_1^1 P_2)$. It follows from (13) that

$$\dot{R} = 2\kappa \text{tr}[P_1^{-1} P_3 (P_1^{-1} P_2)^2],$$

which is non-negative, by Lemma 10.4. The rule (11) (or at least its deterministic version (12)) thus implies that the reduction $A^T y$ improves in performance with time, in that R increases and the reconstruction error decreases.

Moreover, by Lemma 10.4, R can cease to increase only when A is of the form $H_p J$, where the columns of H_p are p distinct eigenvectors of V and the $p \times p$ matrix J is nonsingular. When this point is reached then the reconstruction $\hat{y}$ is just $\sum'_j \xi_j c_j$, where the eigenvectors c_j appearing in the summation $\sum'_j$ are just the columns of H_p. One finds that the only equilibrium of (12) which is stable is that for which these eigenvectors are those corresponding to the p largest eigenvalues, when $\hat{y} = y_{(p)}$. The matrix A will converge to this equilibrium manifold from any starting point except for those in the unstable equilibrium manifolds.   □

Although the reconstruction $y_{(p)}$ delivered by the algorithm is unique, the reduction $x = A^T y$ is not. In fact, one finds (see the note) that the equilibrium solutions of (12) are of the form $A = H_P J$, where H_p has as columns a selection of p distinct eigenvectors of V, as above, and J may be any orthogonal $p \times p$ matrix. The reason for the restriction to orthogonality will emerge. The reduction x thus consists of p distinct principal components (the first p, in fact), but possibly linearly mixed to the extent of a rotation in $\mathbb{R}^p$.

This possible mixing of principal components has been regarded as undesirable: one would like x to have the immediately interpretable form of a listing of the values of the first p principal components. Modifications to the algorithm which achieve this normalisation have been suggested, generally relying on lateral inhibition of network nodes. This work and much else is reviewed in Diamantaras and Kung (1997).

The learning rule (11) can be motivated by the same empirical steepest descent argument which led to rule (2). Step-wise minimisation of the cost function $E[|y - ABy|^2] = E[|y - Ax|^2]$ with respect to A and B leads to the rules

$$\delta B = \kappa A^T (y - Ax) y^T, \quad \delta A = \kappa (y - Ax) x^T \qquad (10.14)$$

respectively. However, the indeterminacy in the reduced-dimension representation $y = ABy$ means that there is no need to revise both A and B. If we take the normalisation $B = A^T$ and use only the second relation (14) then we have the rule (11). That is, one attempts to regress y upon x, but then lets the regression coefficient A determine the rule $x = A^T y$ by which x is formed from y. It is not clear f that this rather arbitrary telescoping of the descent rules works, but Theorem 10.5 tells us that it does. It is this choice of normalisation which reduces the indeterminacy in $A = HJ$ to the extent that J must be orthogonal.

Note

The spectral representation (5) of V can be written $V = H\Lambda H^T$, where H is the orthogonal matrix whose columns are the eigenvectors of V, and Λ is the diagonal matrix of eigenvalues. The equilibrium condition for equation (12) is $VA - A(A^T V A) = 0$. In terms of $C = H^T A$ this can be written $\Lambda C - C(C^T \Lambda C) = 0$. But this implies that the rows of C are either distinct eigenvectors of $C^T \Lambda C$ (with eigenvalues shared with V) or are zero. There must be exactly p of the first, if A is to have rank p, and there can be no more. If we collect these to make a square matrix J then the equation becomes $\Lambda_p J = J(J^T \Lambda_p J)$, where Λ_p is the diagonal matrix of eigenvalues of V which figured in the nontrivial relations. Premultiplication by J^T yields $J^T J = I$, so the square matrix J is orthogonal. We then have $A = HC = H_p J$, which is of the form asserted.

10.5. Dual variables, back-propagation and Hebb's rule

The back-propagation technique for a neural net constitutes an interesting example of supervised learning. The usual formulation of network performance is that it should realise a prescribed input–output relation, in that, if output for a given input y is $\phi(y)$, then one would like the function ϕ to approximate a prescribed function $\bar{\phi}$ as well as possible. A cost function penalising the discrepancy is set up, and the synaptic weight matrix W of the network is modified to minimise this cost, averaged over values of input y.

We shall rather take a statistical view of the problem, in that we shall require the output of the net to approximate a random variable z as well as possible. That is, input y and desired output z are specified as jointly distributed random variables, and one wishes to choose W to make $\phi(y)$ deviate from z as little as possible, in some probabilistic sense. It is by comparison of actual output $\phi(y)$ with desired output z that supervision is exercised.

Let x be the vector of node outputs and y the vector of external inputs to the nodes. Then x is determined by the set of equations

$$x = f(Wx + y). \tag{10.15}$$

We shall suppose that there is a penalty function

$$C(y, z, W) = \frac{1}{2}(x - z)^T G(x - z), \tag{10.16}$$

a non-negative definite quadratic form, which penalises departures of x from an ideal output z. Here the submerged dependence of x upon W has been indicated explicitly.

In actual cases the external input y would be applied only to a subset of the nodes of the net (the *input nodes*) and output taken only from the subset of *output nodes*. However, it is formally simpler to take the general approach of (15), (16). One specialises simply by assuming that y_j is zero if j is not in the input set and that the matrix element G_{jk} is zero if either j or k is not in the output set.

Let us for simplicity assume first the case of a totally linear net, when (15) reduces to

$$x = Wx + y. \tag{10.17}$$

We wish to determine the learning rule derived by gradient minimisation of $E[C(y, z, W)]$ with respect to W, but to do so without actually solving equation (17) for x. Let us then consider the Lagrangian form

$$L(x, y, z, \mu, W) = \frac{1}{2}(x - z)^T G(x - z) + \mu^T(x - Wx - y) \tag{10.18}$$

where μ is a vector of Lagrangian multipliers (the *dual variables*) associated with the constraint (17). By extremising this form with respect to x and μ we deduce the value of the penalty function (16) when the solution x of (17) is substituted into it. The extremal condition with respect to μ is just (17); the extremal condition with respect to x is

$$G(x - z) + \mu - W^T \mu = 0. \tag{10.19}$$

Equations (17) and (19) determine x and μ in terms of the values y and z specified for the current learning trial, and we have then

$$\frac{\partial C}{\partial W} = \frac{\partial L}{\partial W} = -\mu x^T.$$

The empirical improvement rule for the element w_{jk} thus takes the form

$$\delta w_{jk} = \kappa \mu_j x_k, \qquad (10.20)$$

where x and μ are the values for the current trial.

The rule (20) is remarkably simple and intriguing, in that the recommended increment in w_{jk} is a product of a term for node j and a term for node k. In this it recalls the celebrated Hebb's rule, derived from physiological observation, but much invoked in the ANN context. Hebb observed that a particular synaptic connection tended to be reinforced if the two neurons it connected were simultaneously excited; it otherwise tended to decline. The product in (20) represents the notion of simultaneous excitation.

However, the factors in the rule (20) are not as simple as 'node excitation'. The term x_k reflects the information or 'knowledge' gained at node k from knowledge of the input y. The term μ_j reflects the 'need-to-know' at node j determined from the observed discrepancy between actual and ideal outputs. It is then beneficial to increase the information flow from node k to node j if knowledge available at k matches need-to-know at j.

Relation (19) in fact expresses the *back-propagation algorithm* which determines the μ_j. If one regards W as expressing forward propagation through the network then the transpose W^T expresses propagation through the same network in a backward direction. The terms 'forward' and 'backward' become unequivocal if propagation is indeed undirectional, in that the network is layered. That is, if it consists of successive layers, the input layer being the first and the output layer the last, and if w_{jk} can be non-zero only if j lies in the layer following that containing k.

Relation (2) is of course a particular case of (20), with the multiplier μ exhibited as proportional to the discrepancy in output.

If we return to the nonlinear case (15) then the Lagrangian form (18) becomes

$$L(x, y, z, \mu, W) = \frac{1}{2}(x - z)^T G(x - z) + \mu^T [x - f(Wx + y)]. \qquad (10.21)$$

The back-propagation relation (19) and the learning rule (20) then become

$$G(x - z) + \mu - W^T \Gamma \mu = 0$$
$$\delta w_{jk} = \kappa' \gamma_j \mu_j x_k.$$

Here γ_j is the derivative of f at the argument $\sum_k w_{jk} x_k + y_j$, and Γ is the diagonal matrix with elements γ_j.

10.6. Self-organising feature maps

We have worked almost entirely with linear models, which are suggestive and tractable, but plainly unrealistic in many cases. However, it is not clear that there is any more general class of natural models which offers the same degree of guidance. Rather, one has to look for reduction principles which seem natural and general in themselves. One strikingly simple and appealing such concept is that of a self-organising feature map, proposed as an ANN technique by Kohonen (1982, 1989) and as a physiological principle by Amari (1977b, 1980, 1983); see also Takeuchi and Amari (1979) and Linsker (1986, 1988).

The Kohonen algorithm starts from the Hamming net of Section 5.2. Suppose the data input y takes values in a data space $\mathcal{Y}$. The inferential net has m nodes, labelled by j ($= 1, 2, \ldots, m$), and associated with node j is an element a_j of $\mathcal{Y}$. When data y are fed in then that node j is activated for which $D(a_j, y)$ is minimal, where D is a measure of distance between the two arguments.

Up to now this is pure Hamming — exactly as in Section 5.2 if we assume that y is an n-vector and $D(a, y) = |y - a|^2$. The incoming data y is classified abstractly by the node j that it activates; the associated a_j represents a reconstruction of y from the reduced description j. That rule is chosen for which the reconstruction is most faithful, as measured by D.

However, what is now different is that the 'traces' a_j are not prescribed, but are learned on the basis of an organising principle. Suppose that the nodes are regarded as sites in a 'neural space' $\mathcal{N}$, in which node j has coordinate r_j. Let $j(y)$ be the node activated by input y. Then the traces a_k are modified by the learning rule

$$\delta a_k = -\kappa \phi(j(y), k) \frac{\partial D(a_k, y)}{\partial a_k}. \tag{10.22}$$

Here $\phi(j, k)$ is a non-negative weighting, generally symmetric in its arguments, which decreases from 1 to 0 with increasing distance between the two sites in $\mathcal{N}$. It could be chosen, for example, as $\phi(j, k) = \exp[-\kappa' d(j, k)]$, where $d(j, k)$ is the distance between nodes j and k in $\mathcal{N}$, and κ' is a parameter which can be varied.

The rule thus takes neighbourhood relations into account, in that it not merely brings the trace for node $j(y)$ into greater agreement with the observation y, but does the same (in a degree decreasing with distance) for neighbouring nodes as well. The effect of this is that the map ultimately 'organises itself', in the sense that inputs which lie close to each other in $\mathcal{Y}$ are mapped on to (i.e. activate) nodes which are close to each other in $\mathcal{N}$. Here 'distance' is measured by D and d respectively.

The learning rate κ is decreased with time to allow the map to stabilise. The spatial discrimination is also sharpened (by increase of the parameter κ', say) so that modification becomes increasingly local.

The maps which emerge from this simple rule are most striking; several examples are presented in Hassoun (1995). Typically the neural space $\mathcal{N}$ is taken as a subset of the plane (with the page and the cortex in mind). Suppose that $\mathcal{Y}$ is discrete, a set of c colours say. One has a measure D of 'distance' between colours, and d is taken as Euclidean distance in the plane. Then the map will so organise itself that all nodes which respond to a given colour will form a single connected block, with colours which are close being represented by adjacent blocks.

The mappings induced are also most interesting from the purely mathematical point of view. Suppose that $\mathcal{Y}$ and $\mathcal{N}$ are subsets of $\mathbb{R}^n$ and $\mathbb{R}^q$ respectively, and that D and d are the corresponding Euclidean distance functions. For $n = 1$ and $q = 2$ a line-interval in $\mathcal{Y}$ is mapped on to $\mathcal{N}$ as a space-filling curve. In the case $n > q$ the mapping has the impossible task of reconciling reduction of dimension with continuity. It resolves the dilemma by inducing a cellular structure, mapping being continuous in cell interiors and discontinuous at cell boundaries. Takeuchi and Amari (1979) have demonstrated that the columnar nature of the cerebral cortex could be explained in this way.

When it comes to implementation, the question which arose for the Hamming net poses itself again: how is the matching step, i.e. the minimisation of $D(a_j, y)$ with respect to j, to be performed? In fact, by 1980 Amari had thought deeply about these matters in the context of physiological neural architecture, having developed a penetrating mathematical analysis of natural neural nets over several years. In the papers quoted above he concerned himself explicitly with the question of self-organisation, obtaining results which he himself justifiably regards (Amari, 1990) as constituting one of the deeper theories on neural networks.

Amari's model is physiologically based, with dynamics which produce a response automatically localised to the best-matching nodes. It is set in continuous space, so that $\mathcal{Y}$ and $\mathcal{N}$ are connected sets in $\mathbb{R}^n$ and $\mathbb{R}^q$, say. The sites r of $\mathcal{N}$ are regarded as actual neurons (or, rather, neural masses — see Chapter 11). A neuron with 'excitation' u produces an output of rate

$f(u)$, where f is the familiar sigmoid activation function. (In fact, u is a membrane potential and $f(n)$ the rate of the consequent pulse output; see Sections 11.2 and 11.3.) As a function $u(r,t)$ of space and time u obeys the equation

$$\dot{u} = -\alpha u + \beta \circ f(u) + a^\tau y + h. \tag{10.23}$$

The composition term $\beta \circ f(u)$ represents the lateral effect of neighbouring neurons; its explicit form is

$$\beta \circ f(u(r,t)) = \int_{\mathcal{N}} \beta(s) f(u(r-s,t)) ds. \tag{10.24}$$

The weighting function β plays for this model the role that ϕ played for the Kohonen model. However, it is assumed to have the 'on-centre off-surround' character of Fig. 10.1. That is, neurons reinforce others which are close, but inhibit others which are distant. This is a familiar pattern in these contexts; it has the effect of increasing contrast. The input vector y is supposed constant in space and time, as is the bias term h. However, the coefficient a of y is a position-dependent vector $a(r)$. Again, one can regard $a(r)$ as the 'reconstruction' of y which the net would produce if the neuron at r were activated. The dynamics must be such that the neuron $r(y)$ which best matches input y is located, and a form of $a(r)$ must be

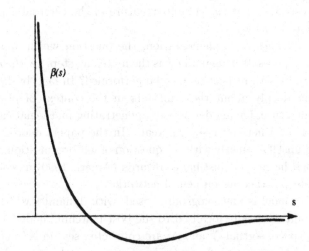

Fig. 10.1. The form of the lateral weighting function $\beta(s)$ of relations (23) and (24), demonstrating 'on-centre off-surround' behaviour required for clear delineation of decision regions.

learned which produces both a mapping $r(y)$ which is mostly smooth and a reconstruction $a(r(y))$ which is tolerably faithful.

The response of the neural system to the input y is regarded as the equilibrium solution $u(r, +\infty)$ of (23); let us write this as $u^*(r, y)$. Then, for given y, the response will be maximal at some value $r(y)$, and Kishimoto and Amari (1979) demonstrate that indeed the response is positive only in a neighbourhood of $r(y)$. The size of this neighbourhood can be controlled by choice of the bias h.

Suggested learning rules are a Hebbian rule

$$\delta a(r) = \kappa[-\lambda a(r) + f(u^*(r, y))y]$$

and the analogue of (22)

$$\delta a(r) = \kappa u^*(r, y)[y - a(r)].$$

The degree of smoothing implied by the operator β in (23) determines the 'graininess' of neural space: the less the smoothing, the more time the net takes to organise itself, but the greater the discrimination ultimately achieved. However, there will be a physiological lower limit to graininess, setting an upper limit on information density in $\mathcal{N}$.

The variables $a(r)$ of the Amari treatment (or a_j of the Kohonen treatment) correspond indeed to the 'simple memory traces' of Chapters 5 and 6. The techniques come near to estimating actual traces in a way that the principal component analysis of Section 3 does not, because they bring the trace evoked by the individual observation y into closer agreement with y itself rather than ask simply for agreement in a linear mean-square sense. The spatial structure used to organise the traces has the effect that a somewhat discrepant y can be accepted as emanating from a variant of the correct trace rather than from the incorrect trace.

However, these techniques deal only in simple traces, and are not able to recognise a composite stimulus as such. That is, the neurons act as 'grandmother cells' in that they recognise a trace as a familiar whole, rather than as possibly a composite of familiar features. The distance function D could indeed be so constructed so as to represent compositeness (in that two trace values are regarded as closer if they share a common feature), but to formulate matters thus is to assume the feature recognition problem already solved. A network capable of learning to recognise both individual features and composites of them will necessarily be more complicated.

10.7. Other proposals

A great part of the ANN literature could be said to be concerned with the topics which we have pursued somewhat further in this chapter: ways of learning to recognise images and to recognise the features of which images are composed. A review is then out of the question.

However, even the most cursory of sketches should mention Grossberg's development of ART: adaptive resonance theory. (See Grossberg (1976a, 1976b, 1982, 1988) and papers quoted there; also Carpenter and Grossberg (1987) for an elaborated version of these ideas.) Grossberg's devices incorporate a number of traces a_j, which he describes as 'templates'. When a new data vector arrives the best-matching template is identified. If the match is good enough then that template is perturbed to slightly improve agreement yet further. If agreement is not satisfactory then y is adopted as a new template. The actual procedure is somewhat more complicated, but this is the essence. The process of backward and forwards comparison is then just that which we have observed in a number of contexts: the reduction of the sensory image to an abstract classification, followed by the reconstruction of a sensory image from this classification and comparison with the actual data y, Grossberg speaks of 'adaptive resonance' having occurred when the reconstruction agrees sufficiently well with the original.

A feature of Grossberg's approach is the use of network dynamics, embodied in systems of differential equations, to standardise the data and to seek the best match. Simple versions of these have already been sketched in Section 3.6. In particular, the match-seeking dynamics are essentially identical with those developed by Amari (see Section 6) — an example of convergent evolution.

The Hopfield network itself has long been held up as an example of unsupervised learning, in the sense that the synaptic matrix $W = AA^T$ can be identified as proportional to $E(yy)^T)$, and so learned by the 'Hebbian' rule $\delta W = \kappa yy^T$ plus a normalisation. The normalisation would bring the rule to something more like $\delta W = \kappa(yy^T - W)$. In fact, $E(yy)^T)$ has the form $AUA^T + V$, where $U = E(rr^T)$ and $V = E(\epsilon\epsilon^T)$. The occurrence of U probably improves matters and the occurrence of V probably does the reverse. However, weightier objections are that this learning rule seems to have no performance-related basis, and that the standard Hopfield net in any case faces the queries listed in Chapter 6.

Dayan and others (Dayan *et al.* (1995), Dayan and Hinton (1996)) have proposed the Helmholtz machine. This is an associative memory, based again upon the sensory/abstract alternation. It learns in an unsupervised fashion in that its criterion is, again, economy of the reduced representation plus faithfulness of the reconstruction. It has a multi-layer design and its operation seems rather complicated, although a motivation is suggested for the learning rules. However, the authors claim that the machine has the capacity to recognise structure in the input — a major advance, if confirmed.

Part III

Oscillatory Operation and the Biological Model

Any comparison between artificial neural networks (ANNs) and biological neural networks (BNNs) must be interesting, because of the infinitely higher level of function, sophistication and economy to which BNNs have successfully evolved. If one sees a difference between the two at some level of mechanistic detail, then one must ask whether the difference represents an advantage achieved by the BNN from which one should learn, or whether it merely represents a way of coping with biological limitations — the neural equivalent of having to find an alternative to the wheel.

The differing operation of BNNs and conventional ANNs has been particularly studied by W.J. Freeman in a sustained and insightful series of papers. In the bibliography we list some of his works between 1964 and 1994, together with papers by or with his co-workers: Baird (1987), Bressler and Freeman (1980), Bressler (1987), Eeckman and Freeman (1991), Eisenberg, Freeman and Burke (1989), Freeman and Schneider (1982) and Freeman, Yao and Burke (1988). These are based on a broad range of studies by himself and his co-workers, covering direct observation of electrical activity in the animal brain, analysis of mathematical models of the structures deduced and experimentation with electronic analogues of these structures. We shall refer to individual papers at the appropriate points, but would particularly note the two survey papers (Freeman, 1992 and 1994), which summarise much of the work and set out Freeman's general theses.

Freeman's starting point is the fact which has been strikingly evident from the time electrical activity of the brain could first be observed: that biological neural systems show continual oscillation of a rather irregular character, even under steady-state conditions (see Fig. 11.1). This is in contrast to the behaviour of an ANN, whose variables converge smoothly to a static equilibrium, representing the solution of the problem the net has been set. In the ANN context significant variables are represented by the absolute magnitudes of the network variables. However, in the biological context the absolute magnitude of a variable is almost meaningless. Information must then be conveyed by patterns of temporal variation rather than by a static setting of level. Observation, as much as considerations of this nature, leads Freeman to contend that BNNs represent the values of significant variables by the dynamic state which the net adopts. Thus, an external stimulus which the BNN has learned to regard as belonging to some significant class will drive the system into the dynamic state associated with that class.

These collective oscillations are manifestations of the fact that even physically separated parts of the neural system are locked into a single mutually-interacting dynamic system, showing continuing activity even under steady-state conditions. One can see the advantages of such an arrangement. First, communication and synchrony over a distance are thereby achieved, with a degree of immunity to noise and to slow irrelevant variations in baseline activity. They are also achieved with a degree of insensitivity to the exact form of the signal, which need not and in general will not be exactly periodic. Second, if the dynamics are based upon a natural extremal principle then the system finds the extremum appropriate to the data it has received by responding to its own geometry rather than by searching and testing. It does this simply by sinking into the basin of attraction in which the external stimulus has placed it, and settling to the corresponding dynamic mode.

So, one should not think, for example, of the olfactory bulb's 'transmitting' information to the olfactory cortex as one would transmit over a communication channel (see Chapter 14). Transmission pathways exist, but so do reverse pathways, and an appropriate sensory stimulus will bring the whole coupled system to a pattern of joint oscillation of a spatial form dependent on the stimulus. It is by this mutual interation of bulb and cortex that a collective dynamical state is created which implies information transfer. The experience of the particular dynamical state constitutes the inference formed from the stimulus, an inference which Freeman contends the organism can make directly.

Any electroencephalogram of the type of Fig. III.1 reveals oscillations indeed, but ones which are so complex and irregular that one may see the system as chaotic. Freeman's electronic analogues show the same behaviour. This must be a consequence of the combined effects of complexity, nonlinearity and delay. However, the point is that the system can detect the *spatial* pattern of variation (i.e. the distribution of amplitude over the nodes of the system) and may even be able to characterise the temporal

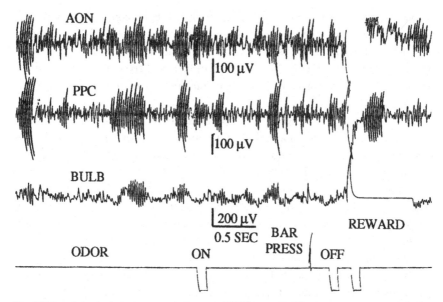

Fig. III.1.   A typical electroencephalogram (EEG record). The traces represent variation at successively higher levels of the olfactory system: the mitral cells, the olfactory bulb, the anterior olfactory nucleus and the prepyriform cortex; see Chapter 14. Reprinted from *Progress in Brain Research*, 102, W.J. Freeman, 'Role of chaotic dynamics in neural plasticity', 319–333 (1994) with the kind permission of the author and of Elsevier Science–NL, Sara Burgerhartstraat 25, 1055 KV Amsterdam, The Netherlands.

variation in some broad statistical sense. In other words, chaotic variation may be no bar, although one may doubt whether it is of direct benefit.

Freeman (1992) speaks of a 'carrier'; see also Skarda and Freeman (1987). By this he means, not something like the sinusoidal carrier used for the one-way transmission of radio signals, but that component of variation which is common to all parts of a BNN such as the olfactory system, and which accounts for about 70% of the variation at a given point. Rather than being sinusoidal this is complex in form, even chaotic. Rather than carrying a one-way transmission, it is evidence of the integration of the system into a dynamic whole by reciprocal links. Rather then being modulated temporally, it is modulated spatially, in that it is the spatial pattern of amplitude which primarily characterises the dynamic state of the system (although there are, of course, temporal consequences).

Freeman carries his thesis rather further when he sees the restlessness of a BNN as evidence of a continuing goal-driven and information-seeking faculty, in contrast to the behaviour of an ANN, which will meekly revert to passive inactivity when its assigned task has been completed. However, this interpretation, when valid, is surely a reflection of level of development. ANNs have hitherto been asked only to solve one strictly delimited problem at a time; the day will come when they will be asked to cope with a dynamic environment and to actively direct their sensors to gain information relevant to the current need.

An investigation of these ideas separates into at least three stages. First, one must set up models for individual neurons and for assemblies of such neurons. There is a very large literature on such matters; in Chapter 11 we give a treatment suited to our purposes.

Then one must consider the particular mechanism which disposes the system to oscillation. Many authors have contributed ideas, but none (in my view) more cogently than Freeman, who has combined examination of and experiment on biological material with mathematical and analogue investigation of models. We describe Freeman's basic ideas in the latter part of Chapter 11 and pursue their implications for systems in Chapter 12. Some more general discussion of the literature is given in Section 12.10.

Finally, one must develop an associative memory based on these mechanisms. Guiding principles are needed, and it was for this reason that we devoted most of Part II to the development of guiding principles for the ANN-realisation of an associative memory. That line of thought led us to the probability-maximising algorithm, and it is on to the PMA that we graft Freeman's oscillatory mechanisms in Chapter 13.

However, the truly final stage must be comparison with physiological reality. In Chapter 14 we compare the systems deduced with those evident in the olfactory system, which is for various reasons a good case to examine. In that chapter we claim a striking correspondence between the two, and are indeed emboldened to make testable predictions.

Chapter 11

Neuron Models and Neural Masses

11.1. The biological neuron

If we are are to consider the operation of BNNs, with a view to perceiving features which could be advantageously incorporated in ANNs, then we should start with at least a brief description of the unit from which they are formed: the biological neuron. Some features of this neuron may have a general significance; others may reflect only the limitations imposed by biological realisation. One should not try to decide which is which prematurely.

The neuron is described in many texts and at many levels of detail. The description given in Freeman (1992) conveys the essentials very clearly.

We sketch the basic neuron in Fig. 11.1. The cell body receives inputs (as many as 100,000, from other neurons) through a set of conductors, the *dendrites*, which form a branching body, the *dendritic tree*. It sends a single output through the *axon*, a long conductor which branches when it reaches its target area: the set of neurons to which it provides input.

Input signals evoke an electrical response in the dendrites. This has the advantage that the electric currents from the individual dendrites can be summed in the cell body, to form a net signal (the *cell current*, which then induces the *membrane potential*) which provides the basis for the neuron output. However, it has the disadvantage that it is intrinsically a high-loss form of transmission. For this reason, the output is transmitted through the axon in electrochemical pulses, of a standard magnitude but at a rate increasing with the membrane potential. The axon is so constructed that losses are made good in passage, and that transmission is as rapid as an electrochemical structure could achieve. It is then the axon which carries signals over distances; the dendrites merely provide a means of multiple input in the immediate neighbourhood of the neuron.

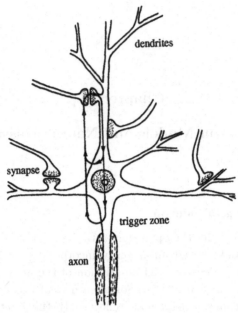

Fig. 11.1.   A schematic drawing of a neuron. The dendrites are shown radiating upwardly and laterally from the cell body; they meet incoming axons at the *synapses*. Each synapse acts as a small battery with the effect, for an excitatory synapse, that a current is driven from the synapse through the cell body. This current is indicated by the loop in the diagram; it leaks out into the interneural medium, principally at the *trigger zone*, and then completes its circuit by return to the other side of the synapse. For an inhibitory synapse the flow is in the reverse directions. Currents from different synapses sum in the cell body to the *cell current*, which induces a build-up of charge on the membrane at the trigger zone. When the potential associated with this charge reaches a threshold level then the cell 'fires', with the effect that the potential subsides and an electrochemical pulse is transmitted out through the axon at the bottom of the diagram. Adapted from *Progress in Brain Research*, 102, W.J. Freeman, 'Role of chaotic dynamics in neural plasticity', 319–333 (1994) with the kind permission of the author and of Elsevier Science–NI, Sara Burgerhartstraat 25, 1055 KV Amsterdam, The Netherlands.

There is then a necessity to transform between the two representations of a signal: in terms of a variable pulse density and in terms of a variable potential. Incoming pulses are converted to an electric current at the *synapses*, the junctions between incoming axons and the dendrites of the target neuron. As with every other aspect of the neuron, the synapse has its own intricate and ingenious structure, However, the net effect is that the synapse behaves like a high-impedance electric battery, in which each incoming pulse generates a momentary *action potential*. The current generated by these potentials travels through the cell body towards the root of the axon. At all points during its passage it tends to leak out of the neuron into the surrounding medium, where it ultimately closes its circuit (and so forms a *loop current*) by returning to the synapses. However, there is a preferential tendency to leak through the membrane of the *initial segment* of the axon — it is to this we refer when we use the term 'membrane'. This initial segment is often referred to as the *trigger zone*, because it is there that excitation of the membrane initiates pulses down the axon.

In fact, these loop currents can flow in either direction: *excitatory synapses* give a current flowing as indicated in Fig. 11.1; *inhibitory synapses* give a current in the opposite direction. The cell current is then a sum of these positive and negative contributions. What EEG sensors detect is a somewhat aggregated quantity: the difference in potential between two 'points' in the interneural medium at which the electrodes are placed. This is roughly proportional to the current between those points, a current to which the return currents of neighbouring neurons all contribute.

It is at the initial segment of the axon, the trigger zone, that the electric current generated at the synapses is converted back into a train of pulses, transmitted onward through the axon. The membrane itself behaves like a leaky capacitor (i.e. a capacitor shunted by a resistance), and builds up a charge, and so a potential, which is a weighted integral of past current through the membrane. What comes into play next is another subtle and powerful mechanism: the 'sodium pump', represented mathematically by the Hodgkin-Huxley model of axonal transmission. Its effect is that, when the membrane potential increases to a critical level, the neuron 'fires', in that an electrochemical pulse is transmitted down the axon and the membrane potential falls to a rest value. No further firing can take place until the membrane potential again rises to its critical value; this enforced rest phase is termed the *refractory period*. The greater the cell current, the shorter this refractory period will be, but there is a positive value below which it cannot fall.

One now sees the plausibility of the linear gate of Chapter 3 as a skeletal model of a neuron. The linear combination of inputs corresponds to the summation of dendritic currents; the sigmoid activation function corresponds to the nonlinear response of output pulse density to membrane potential. However, the linear gate model plainly neglects the internal dynamics of the neuron. Moreover, it should be seen as a static model of an *assembly* of neurons (a *neural mass*) rather than of an individual neuron. The situation is well set out in Freeman (1992), p. 463. The response of membrane potential to input pulse density is somewhat sigmoidal for the individual neuron; it is effectively linear for the neural mass. The response of output pulse density to membrane potential is non-sigmoidal for the individual neuron, in that response is zero below a certain threshold value, initially linear above that value, but then abruptly zero at the *hyperpolarisation* point. However, for the neural mass it is sigmoidal, for reasons explained in Section 3.

This brief account necessarily ignores both detail and the inevitable exceptions and skates over fundamental issues. We should amplify one point: neurons tend to be either excitatory or inhibitory, in that they drive synapses of the one type or the other. *Duke's law* states the biological observation that there is indeed a division between the two roles, and that no neuron drives synapses of both types. There is a third type of neuron, the *modulatory neuron*, which modifies the gain factors of neighbouring neurons. These can, for example, induce *arousal* in a neural mass. They obviously work on a slower time scale than do the millisecond dynamics of other neurons. There are also synapses which are *neuromodulators*; when stimulated these do not produce dendritic current, but rather modify the behaviour of neighbouring synapses. One should also note the existence of *interneurons*; these are inhibitory neurons which often possess no axon, but affect other neurons in the immediate neighbourhood by direct electrical action.

As stated, the dendritic tree is a very rich one, receiving input at 10^3–10^5 synapses. The human cerebral cortex contains of the order of 10^{10} neurons; the whole nervous system of the order of 10^{11}.

11.2. Neural masses: Reduced dynamics

The individual biological neuron is in most cases stochastic in its operation, and cannot carry a significant information flow. To achieve a significant flow one must then have a substantial number N of such neurons operating in parallel: a *neural mass*. We assume the neurons statistically identical and also, for the moment, that they do not interact. They are thus coupled only by their common input, and their averaged output represents the mean response of a single neuron to that input. Such a mass represents the lowest possible level of organisation; Freeman (1975, 1987) denotes such an assembly of non-interacting neurons by K0 in a hierarchy of neural systems.

It is such a neural mass which must be regarded as the basic component of a neural system. Its dynamic behaviour should properly be deduced by a probabilistic analysis of the model for the case of finite N, followed by an asymptotic treatment for the case of large N. We do in fact give the deterministic limit of such a treatment in the next section, but there have been a number of attempts to reach conclusions on the basis of a much reduced model. The first such was given by Wilson and Cowan (1972); it has survived well, and is presented essentially unchanged in a book as modern and erudite as Beckerman's (1997). However, the approximations are Draconian in the extreme, and we shall see that the more detailed analysis of the next section leads us to conclusions which retain essential aspects of the neural dynamics, and whose form is in fact simpler.

Wilson and Cowan use the term 'localised population' rather than 'mass', and the term 'population' duly reminds us that we are indeed dealing with a collection of individual neurons. Let $y(t)dt$ be the proportion of neurons which fire in the infinitesimal time-interval $(t, t + dt)$, so that y can be regarded as the *pulse density* of the output of the mass (more strictly, pulse density per neuron). Let $u(t)$ be the pulse density of the input (per neuron), and suppose that the consequent cell current (averaged over neurons of the mass) is

$$x(t) = \int_0^\infty b(\tau)u(t - \tau)d\tau = \mathcal{B}u(t). \tag{11.1}$$

Here $b(\tau)$ is the response function of current to input after a time τ. It is determined by the dynamics of the axons which carry the input and of the synapses which convert the pulse input to a current. It is this current which is detected in EEG studies and which integrates to produce a membrane potential. We use $\mathcal{B}$ to denote the operator (linear and time-invariant) expressed by the integral in (1).

In relation (1) x is scalar-valued, but u would be a q-vector if there were q distinct inputs to the mass. We shall for the moment assume $q = 1$ for simplicity; this would be essentially the case if the different inputs arrived by dynamically similar axonal pathways and were linearly combined in a common fashion by the statistically identical neurons.

Suppose that the neurons have a refractory period of length δ, and that neurons which are sensitised (i.e. out of their refractory period) will fire in response to excitation x with probability intensity $f(x)$. (That is, the probability that a neuron of excitation x will fire in a time interval of infinitesimal length dt is $f(x)dt + o(dt)$.) The function $f(x)$ will emerge as the familiar sigmoid activation function.

The proportion of the neurons which are sensitive at time t can be written $1 - \int_0^\delta y(t - \tau)d\tau$; one might then plausibly assume the dynamic relation

$$y(t) = \left[1 - \int_0^\delta y(t - \tau)d\tau \right] f(\mathcal{B}u(t)), \tag{11.2}$$

on the basis that both sides represent a firing rate per unit time and per neuron at time t. However, the right-hand member of (2) does not bear this interpretation exactly. In multiplying its two factors one has made an implicit assumption: that the level of excitation in the neuron mass is statistically independent of the number of neurons which are sensitive. There is in fact a statistical relation between the two of a rather unevident form; the analysis of the next section throws some light on this. However, if one neglects this effect and accepts the defining interpretation of $f(x)$ then equation (2) has a degree of plausibility.

The relation which Wilson and Cowan proposed was in fact

$$y(t + \delta) = \left[1 - \int_0^\delta y(t - \tau)d\tau \right] f(\mathcal{B}u(t)). \tag{11.3}$$

In their interpretation $\mathcal{B}u$ is a measure of excitation level (membrane potential) rather than of cell current. They assume that any sensitised neuron will fire the moment excitation exceeds its threshold, and that this threshold varies over the population, $f(x)$ being the proportion of the population with threshold below x. The reason why $y(t)$ was replaced by $y(t + \delta)$ is not clear. In any case, by adoption of a time-smoothed version $\bar{y}(t)$ of $y(t)$ and by Procrustean approximation they reduced relation (3) to the differential equation

$$\delta \frac{d\bar{y}}{dt} = -\bar{y} + (1 - r\bar{y})f(k\bar{u}) \tag{11.4}$$

where r and k are constants. The effect of the approximation is then to lose the dynamics expressed by the operator $\mathcal{B}$ almost entirely.

In fact, it is not clear where the sigmoid function comes from on any interpretation of the argument. The model is bound to seem arbitrary and its analysis confused, because it neglects an essential element: the spread of state-values of the neurons in the mass. The neurons are distributed over their various possible states at any given time, and it is only by looking at the dynamics of this distribution that we can attain some clarity.

We consider these more detailed dynamics in the next section. The outcome is, however, a very transparent one. To a good approximation the input/output relation takes the simple form

$$y(t) = f(\mathcal{B}u(t)), \tag{11.5}$$

where f is a sigmoid function whose form is deduced. The dynamics of axon and synapse are represented by the operator $\mathcal{B}$, the dynamics of membrane and population by the sigmoid function f.

Relation (5) is essentially that assumed by Freeman (1975, 1987) in his long-running study of these matters. In particular, his time-series analysis of EEG records led him to conclude that the input/current relation $x = \mathcal{B}u$ can be seen as the solution of differential equation $\mathcal{L}x = u$ in x. Here $\mathcal{L}$ is a linear constant-coefficient differential operator, in fact of second order, and u appears as an input driving the equation. The relation between pulse-rate input u and pulse-rate output y can thus be expressed by the pair of equations

$$\mathcal{L}x = u, \quad y = f(x), \tag{11.6}$$

the intermediate variable x being identifiable as the net current through the cell body. (Freeman identifies it as membrane potential, and the two are not very different for fast-reacting cells, but we take it as current, for reasons explained at the end of Section 3.) We could write $\mathcal{L}$ as a second-degree polynomial $L(\mathcal{D})$ in the differential operator $\mathcal{D} = d/dt$. Freeman's determination of L from the data indicates that the zeros of the polynomial $L(s)$ are real and negative. That is, the response of x to u is non-oscillatory and stable.

The form of relations (6) is pleasing on many counts. First: it is simple. Second: the internal dynamics of the neuron are well retained in the differential equation. Third: the sigmoidal response of y to x now has a good explanation, as we hope to persuade the reader in the next section. Finally: one has the choice of variables x or y with which to characterise the macro-state of the neural mass — a freedom we have already found useful in Section 3.4.

11.3. Neural masses: Full dynamics and the tank/sump model

As Freeman (1992) emphasises, it is important to understand that the sigmoid response $f(x)$ emerges as a property of the mass. One can see this only by considering the dynamics of the distribution of the neurons in the mass over their possible states. In this section we study a deterministic

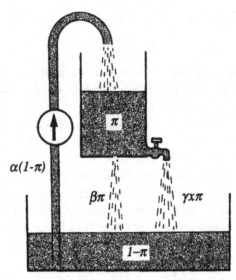

Fig. 11.2. The tank/sump model for the flow of neurons between their two states. The proportion of neurons which are 'sensitised' is in the tank, which they can leave either by simple loss of excitation (as though by leakage at rate $\beta\pi$) or by firing (as though by flow through the tap at rate $\gamma x\pi$, where the cell current x corresponds to the amount by which the tap is opened). The proportion $1 - \pi$, of neurons which are 'refractory', is in the sump, which they leave at rate $\alpha(l - \pi)$ (as though lifted by a pump at a rate proportional to the number in the sump).

version of such a fuller model — essentially, a parametrically-driven compartmental model.

Consider first a model in which the neurons can be in just two states: sensitised or refractory. We shall call this the *tank/sump model*, in that we shall imagine the neuron population as a fluid of unit total amount which is circulating between a tank (the set of sensitised neurons) and a sump (the set of refractory neurons); see Fig. 11.2. Suppose that a proportion π of the neurons is sensitised, so that a proportion $1 - \pi$ is refractory. In what follows α, β and γ are constants. We shall suppose that there is a flow of rate $\alpha(l - \pi)$ from sump to tank; this is the rate per neuron at which refractory neurons recover. This flow is counterbalanced by a flow of rate $(\beta + \gamma x)\pi$ from tank to sump. The component of flow $\beta\pi$ represents those sensitised neurons whose excitation decays before they manage to fire — this is 'leakage' from tank to sump. The component of flow $\gamma x\pi$ represents those sensitised neurons which are stimulated by the dendritic current x to actually fire (and then become refractory) — this is flow from tank to sump through a 'tap' which is opened by an amount x.

The proportion π then obeys the differential equation

$$\dot{\pi} = \alpha(1 - \pi) - (\beta + \gamma x)\pi. \tag{11.7}$$

If x is held constant and equal to $\bar{x}$ then π will converge to the equilibrium value

$$\bar{\pi} = \frac{\alpha}{\alpha + \beta + \gamma\bar{x}} \tag{11.8}$$

and the firing rate in equilibrium is

$$\bar{y} = \gamma\bar{x}\bar{\pi} = \frac{\alpha\gamma\bar{x}}{\alpha + \beta + \gamma\bar{x}} = f(\bar{x}), \tag{11.9}$$

say. We would regard $f(\bar{x})$ as the output pulse rate; note that it is indeed a sigmoid function of the current $\bar{x}$.

Suppose that x shows a small variation $\xi(t)$ about $\bar{x}$; let the corresponding variations of π and y about their equilibrium values be denoted $\zeta(t)$ and $\eta(t)$ respectively. Then the linearised versions of equation (7) and the relation $y = \gamma x \pi$ are

$$\dot{\zeta} = -(\alpha + \beta + \gamma\bar{x})\zeta - \gamma\bar{\pi}\xi, \quad \eta = \gamma(\bar{x}\zeta + \bar{\pi}\xi).$$

We thus deduce that the transfer function from ξ to η is

$$\left(\frac{\eta}{\xi}\right)(s) = \gamma\bar{\pi}\left[\frac{s + \alpha + \beta}{s + \alpha + \beta + \bar{x}\gamma}\right], \tag{11.10}$$

(see the note). This of course only characterises the dynamics for small perturbations, but it is indicative. The response function increases from $\gamma\bar{\pi}(\alpha + \beta)(\alpha + \beta + \bar{x}\gamma)^{-1}$ to $\gamma\bar{x}$ as $|s|$ increases from 0 to ∞, and is to that extent fairly constant. In the transformation from ξ to η there is some attenuation of low frequencies, but the high-frequency response is flat. The reason is that y is proportional to $x\pi$, so that, if π were constant, then y would be simply proportional to x, and the frequency response would be completely flat. If x is changing rapidly relative to π then the response virtually has this instantaneous proportional form. The response is spongier for slower variation of x, because π then has time to change. However, for rapid x-variation one does not do badly by taking the x/y relation simply as $y = f(x)$.

We should increase the number of states to reflect the facts that the refractory period is relatively constant and that there is a range of excitation for sensitised neurons. So, a newly-refractory neuron must progress through a succession of sump states before it becomes sensitised, and can then progress stochastically through the sensitised states, sometimes falling back and sometimes rising to the point of firing. Let this succession of states be labelled $j = 0, 1, 2, \ldots$, let π_j be the proportion of neurons in state j and let π be the column vector with elements $(\pi_1, \pi_2, \ldots)$. (As above, we leave π_0 to be determined by the normalisation $\sum_j \pi_j = 1$, essentially so that we can exhibit $\bar{x}$ as the solution of a linear equation system rather than as the eigenvector of a matrix.) equation (7) then generalises to

$$\dot{\pi} = \alpha(1 - 1^T \pi) - (\beta + x\gamma)\pi, \qquad (11.11)$$

where a is a vector, β and γ are matrices, and 1 is a vector of units. The sump states will be such that one progresses up through them at a constant rate independent of x; the tank states will be such that a neuron can both rise and fall through them spontaneously, but that the presence of current x will encourage rise and increase the rate of firing. The forms of α, β and γ must be such as to reflect these properties.

The equilibrium solution analogous to (8) is

$$\bar{\pi} = (\alpha 1^T + \beta + \bar{x}\gamma)^{-1}\alpha \qquad (11.12)$$

and the firing rate in equilibrium is given by $\bar{y} = f(\bar{x})$, where

$$f(\bar{x}) = \bar{x} 1^T \gamma \bar{\pi} = \bar{x} 1^T \gamma (\alpha 1^T + \beta + \bar{x}\gamma)^{-1}\alpha. \qquad (11.13)$$

Relation (13) generalises (9), and is again a sigmoid function.

Expression (10) for the transfer function from x to y for small variations now becomes

$$\left(\frac{\eta}{\xi}\right)(s) = 1^T(sI + \alpha 1^T + \beta)(sI + \alpha 1^T + \beta + \bar{x}\gamma)^{-1}\gamma\bar{\pi}.$$

As before, the real point is that this expression departs from a constant only at low frequencies, indicating a degree of attenuation there, so that we can represent the x/y relation well enough by $y = f(x)$, with f determined by (13).

Figure 11.3 illustrates the form that one expects for the equilibrium distribution over states at low and high levels of dendritic current respectively. In both cases the distribution over sump states is fairly flat, while that over tank states can show a maximum. In the low-x case this maximum represents a balance between ascent and descent through the tank states (with of course some loss by firing towards the top of the range). For the high-x case there are *fewer* neurons in the tank states (see (8)), because the increased rate of firing depletes them faster. However, this loss in numbers is more than offset by the fact that the neurons are encouraged by the high

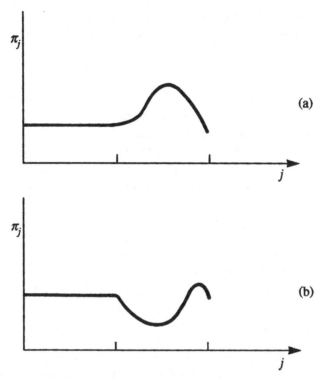

Fig. 11.3. The equilibrium distribution over excitation states: for a low charging rate x in case (a) and for a high charging rate in case (b). The distribution over refractory states is flat, but that over sensitised states shows a peak which moves to the right (despite diminished numbers in these states) as the charging rate increases.

value of x to rush up faster through the tank states; hence the displacement in the maximum.

Freeman ((1994), p. 329; see also (1992), p. 461) makes the point that the neurons spend most of their time at a state just below the point of firing, with relatively few firing at any given time. This is indeed the situation represented in Fig. 11.3. The maximum represents an accumulation of sensitised neurons, almost ready to fire; these are lifted nearer to the point of firing as x increases.

We have taken x as representing the cell current, the charging rate of the membrane, rather than the membrane potential itself. The membrane potential (which is the indicator of degree of excitation of the individual neuron) is represented in this model by the state of excitation j. So, the model does not average membrane potential over neurons. To do this would lose the crucial effect: that individual neurons fire only when sufficiently excited, and that there is a distribution of degree of excitation at any given time. In the model it is the cell current, the charging rate of the membranes, which is averaged.

Note

If a linear relationship $\eta(t) = \int_0^\infty b(\tau)\xi(t - \tau)d\tau$ holds then the transfer function from ξ to η is the Laplace transform $B(s) = \int_0^\infty e^{-s\tau}b(\tau)d\tau$ of the transient response. In frequency terms, an input $\xi(t) = e^{iwt}$ corresponds to an output $\eta(t) = B(iw)e^{iwt}$, transients and possible instabilities apart. However, the behaviour of $B(s)$ as a function of the complex scalar s provides information also on these transients and instabilities.

11.4. Systems of neural units (masses)

It is the neural mass rather than the neuron which is the analogue of the artificial neuron; we shall often term it a 'neural unit' or simply a 'unit' as a convenient term which keeps distinctions clear. The real interest is in networks of such units.

Consider a number of units, the jth having current x_j, in general a function of time $x_j(t)$. Let x be the column vector (x_j), so that the vector of corresponding outputs is $y = f(x)$. Suppose that the input to unit j is $u_j + \sum_k w_{jk}y_k$, where u_j is an external input, again in general time-dependent. This input is a pulse-rate. The synaptic coefficient w_{jk} will be positive or negative according as unit k excites or inhibits unit j.

We assume that all units obey the dynamics (6); the vector x will then obey the set of differential equations

$$\mathcal{L}x = Wf(x) + u, \tag{11.14}$$

where W is the synaptic matrix (w_{jk}).

In (14) we see the version of the McCulloch-Pitts net described in Section 3.4, so ANNs and BNNs have to this extent been reconciled. We shall investigate nets of this type, among others, in the next chapter. However, there is one in particular which we shall discuss now: the Freeman oscillator.

11.5. The Freeman oscillator

Wilson and Cowan suggested in their classic paper (1972) that neural oscillations might be produced by mutual interaction between an excitatory and an inhibitory unit: a biological and nonlinear version of the simple harmonic oscillator. There is plenty of observational support for such a mechanism; it is common to find groups of mitral cells and of granule cells in close association (see Chapter 14), and these are respectively excitatory and inhibitory in their action.

Freeman (1975, 1987) put flesh on this hypothesis. He took the more realistic dynamics (6) of individual masses, showed how the properties of the excitor/inhibitor pair depended upon the character of the operator $\mathcal{L} = L(\mathcal{D})$, and studied the onset of oscillation as parameters were varied. In particular, an increase in (static) input to the system can shift the equilibrium point to a steeper part of the sigmoid response curves and take the system into an oscillatory regime. In this way, a unit built from one or two soft threshold functions can show a sharp threshold effect. For these reasons, and because Freeman has made this two-unit system a fundamental building block for his models, we shall call it the *Freeman oscillator*.

Denote the cell currents of the excitor and inhibitor by x and z respectively. Then the general equation (14) reduces to

$$\mathcal{L}x = -f_2(z) + u, \quad \mathcal{L}z = f_1(x) \tag{11.15}$$

where f_1 and f_2 are the activation functions of excitor and inhibitor, not necessarily assumed identical. Coefficients which one might have included have been absorbed into the definitions of these functions and it is assumed that the only input is a scalar input u to the excitor. Let us assume of $\mathcal{L}$ for the moment only that it has a stable inverse, in that the zeros of $L(s)$ lie strictly in the left half of the complex plane.

Denote the equilibrium of system (15) for a prescribed constant value of u by $(\bar{x}, \bar{z})$. Such an equilibrium exists and is unique; see note 1. Perturbations ξ and ζ of x and z from these equilibrium values obey the linearised equations

$$\mathcal{L}\xi = -\gamma_2\zeta, \quad \mathcal{L}\zeta = \gamma_1\xi, \tag{11.16}$$

where $\gamma_1 = f_1'(\bar{x})$ and $\gamma_2 = f_2'(\bar{z})$ are intrinsically positive. If we look for solutions which are exponential in time, and so proportional to e^{st}, say, then we find from (16) that s must obey the *characteristic equation*

$$L(s)^2 + \lambda^2 = 0, \tag{11.17}$$

where $\lambda = \sqrt{\gamma_1\gamma_2}$. This reduces to $L(s) = \pm i\lambda$.

We can take λ as a coupling parameter. For $\lambda = 0$ the s-roots of (17) lie in the left half of the complex plane, by hypothesis, and the equilibrium point $(\bar{x}, \bar{z})$ is stable. One can imagine that stability weakens as λ increases, revealed by the fact that at least one s-root of (17) moves to the right. If ever a root actually crosses the imaginary axis then the equilibrium $(\bar{x}, \bar{z})$ has become unstable. Suppose that at some critical value of λ such a crossing impends in that equation (17) develops a root on the imaginary axis, at $\pm i\omega$, say. This critical value and the corresponding ω are determined by

$$L(i\omega) = i\lambda. \tag{11.18}$$

The other sign option merely changes the sign of ω.

If at this critical point we find that $\omega = 0$, then that is an indication that a non-oscillatory instability is developing. In such a case the equilibrium is not the centre of a basin of attraction (even in the dynamic sense: that the variable continues wandering but captured in a neighbourhood of the equilibrium). If ω is non-zero, then this is an indication of an oscillatory instability in the linear approximation. In many cases such a transition marks a *Hopf bifurcation*, for which the oscillatory instability in the linearised model marks emergence of a *limit cycle* around the equilibrium point for the actual nonlinear model. Conditions are needed for this to be the case (see Section 4.1), which are in one way or another satisfied in our contexts. The transition is described in more detail below. We shall refer to the value of ω at this transition as the *threshold frequency*.

In the oscillation which develops at criticality the excitor leads the inhibitor by a quarter-cycle, just as velocity leads displacement for the simple harmonic oscillator (see note 2). This predicted phase lead is confirmed by physiological observation. According to Eeckman and Freeman (1990), it has been found in every cortical region in which it has been sought.

Consider the first-order case, $L(s) = s + L_0$, where (by the stability assumption) $L_0 > 0$. Then the roots of (17) are $s = -L_0 \pm i\lambda$. These lie strictly in the left half-plane, whatever λ, and the equilibrium point is always stable. The equilibrium is approached by a damped oscillation whose frequency increases with λ but whose damping remains constant.

Consider the second-order case, $L(s) = s^2 + L_1 s + L_0$, where the assumption of stability would imply that L_0 and L_1 were both positive, and the additional assumption of real zeros (non-oscillatory response) would imply that $L_1 > 2\sqrt{L_0}$. We find then from (18) that $\omega = \sqrt{L_0}$, and that the critical value of λ is $L_1\omega = L_1\sqrt{L_0}$. That is, if the equation

$L(\mathcal{D})\xi = 0$ is regarded as the equation of a lone damped oscillator, then the excitor/inhibitor system begins to oscillate when the coupling exactly cancels the damping, and at a frequency which is exactly that which the lone oscillator would have if undamped.

All these observations are due to Freeman (1975, 1987), as is the experimental determination of the nature of $\mathcal{L}$. We shall henceforth assume $\mathcal{L}$ to be second-order, and shall write $L(s) = s^2 + cs + d^2 = (s + \alpha)(s + \beta)$, where a and β are real and strictly positive. We shall give a more exact discussion of criticality for this case in Section 12.1.

Recall that a frequency of ω radians per second corresponds to one $\omega/2\pi$ cycles per second, or $\omega/2\pi$ *hertz*, and to a period of $2\pi/\omega$ seconds. Freeman, Yao and Burke (1988) evaluate the time constants of the response of current to input for the rabbit as 1.4 and 4.5 milliseconds, meaning that α and β take the values $1/1.4 \approx 0.714$ and $1/4.5 \approx 0.222$ if we take the millisecond as our time unit. To two-figure accuracy this implies that $c = 0.94$ and $d^2 = 0.16$. The threshold frequency is thus $\omega = d = 0.4$ radians per millisecond, corresponding to a period of $2\pi/d \approx 15.71$ milliseconds. In hertz (cycles per second) this would be

$$\frac{1000d}{2\pi} = \frac{400}{2\pi} \approx 63\,\text{Hz}.$$

We shall see in Section 7 that this lies quite centrally in what is known as the gamma band: a range of frequencies which are prominent in EEG records. Activity in this band is actually strongest round about 40 Hz, but we shall see in the next chapter that the threshold frequency drops when oscillators are linked into a system.

In his hierarchy of neural systems Freeman denotes a system (14) which includes both excitatory and inhibitory units by KII. This then properly includes both the Freeman oscillator and what we shall in Chapter 12 term a *bank* of Freeman oscillators. We shall reserve the notation for this latter case.

There is of course a very large literature on systems of neural oscillators; we compare some recent approaches to that of the Freeman school in Section 12.10.

Notes

1. Suppose that $L(0)$ is strictly positive. If we look for an equilibrium solution of the system (15) then it follows from the second equation that z is a monotone increasing function of z. If we substitute this

expression for z into the first equation then we find x equated to a monotone unbounded decreasing function of x. Such an equation has exactly one solution.

2. Suppose that at criticality $\xi = a_1 e^{i\omega t}$, $\zeta = a_2 e^{i\omega t}$. Show that $a_1/a_2 = i\sqrt{\gamma_2/\gamma_1}$, and hence that the excitor leads the inhibitor by a quarter-cycle.

3. At the transition point for the Freeman oscillator equation (17) has solutions $s = \pm id$ and $s = -c \pm id$. This second pair then corresponds to a strictly stable mode.

4. Consider the system $\mathcal{L}x = -\lambda \text{sgn}(z)$, $\mathcal{L}z = \lambda \text{sgn}(x)$, a symmetric Freeman oscillator with hard threshold. Suppose that $\mathcal{L}$ has the second-order form used by Freeman. Show that there is a periodic solution (for any positive λ) in which the two variables follow square waves, x leading z by a quarter-cycle, with period T determined by $\alpha(l + p^2)(l - q)^2 = \beta(l + q^2)(l - p)^2$, where $p = \exp(\alpha T/4)$ and $q = \exp(-\beta T/4)$. In the confluent case $\alpha = \beta$ this has the approximate solution $T \approx 6.074/\alpha$, whereas the linear theory would have given $T = 2\pi/\alpha \approx 6.284/\alpha$. In the case when α is fixed and β is allowed to become small one finds that $T \approx 4/\sqrt{\alpha\beta}$.

11.6. Numerical solution for the Freeman oscillator

One would of course like the full nonlinear treatment of the system (15). We can give the some numerical solutions for what will henceforth be taken as the standard case:

$$L(s) = s^2 + cs + d^2 = s^2 + 0.94s + 0.16. \tag{11.19}$$

We shall assume that

$$f_1(x) = f(x) = e^z(1 + \epsilon^z)^{-1}, \quad f_2(z) = z.$$

The logistic form for f_1 is a standard and convenient one. The simple linear form for f_2 is motivated by an observation made in the literature (see e.g. Baird (1986), Eeckman and Freeman (1990)): that local inhibition is often achieved by a direct electrical effect rather than by mediation through membrane, axon and synapse.

The linearised treatment of Section 5 leads to the conclusion that oscillation should begin when u increases to about 0.79, and should then have a period of 15.71 milliseconds. This is borne out by the numerical solution. There is a limit cycle in all cases for which the static equilibrium

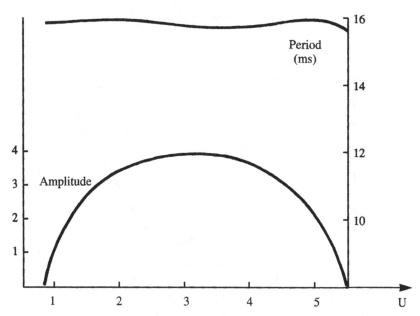

Fig. 11.4. A plot of amplitude and period of the limit cycle (of the excitor current x) for the Freeman oscillator, as functions of the static input u which determines the operating point. Oscillation breaks out at the predicted value $u = 0.79$ and with the predicted period of about 15.7 milliseconds. As u increases further the amplitude (lower graph) first increases and then decreases, but the period (upper graph) is remarkably stable.

is unstable. In Fig. 11.4 we give a plot of its amplitude and period against u. As u increases, the amplitude increases from zero for u just below 0.8 to a maximum at u about 3.5 and then declines, reaching zero again for u about 5.5. This is as one would expect: as u increases so does the static equilibrium value of x, and the slope of f at this value first increases and then declines. One would not expect this upper saturation region to be reached under normal operation.

The period shows remarkably little dependence upon the value of u, varying between 15.71 and 16. It increases as the curvature of $f(x)$ at the operating point $\bar{x}$ increases in magnitude, but the effect is small.

Figure 11.5 shows the variation of x with time for $u = 1.5$; it is very close to sinusoidal. This is borne out by the x/z plots for varying u given in Fig. 11.6. These are close to elliptic in form, which is at least consistent with joint sinusoidal variation. As one would expect, the operating point (the rough centre of the ellipse) increases in both coordinates with increasing u, but the amplitude of oscillation first increases and then decreases.

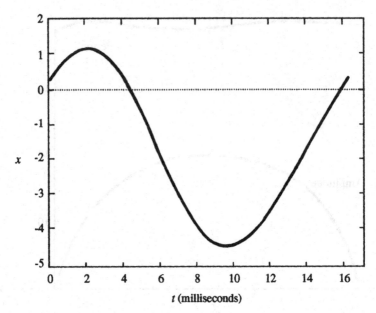

Fig. 11.5. A plot of $x(t)$ for the Freeman oscillator with $u = 1.5$. The variation is very close to sinusoidal.

11.7. The evidence of EEG traces

EEG traces such as that depicted in Fig. III.1 provide the most accessible observations on neural activity. Recall that these measure the potential difference between the sensor points. Freeman (1994) refers to this potential as the 'extracellular compound post-synaptic potential', emphasising that it reflects the net effect of the flow of return currents in the interneural medium. What is measured is then effectively the current between the sensor points: a sum of corresponding components of return current for neurons in the neighbourhood.

There are recognised bands of frequencies which are prominent under some of the conditions of waking, sleeping, attention, etc. These are the delta band of 1–3 Hz, the theta band of 4–7 Hz, the alpha band of 8–12 Hz, the beta band of 15–25 Hz and the gamma band of 25–90 Hz. These are broadly observed across mammalian species; Bressler and Freeman (1980) quote these figures for cat, rabbit and rat, and the human picture is not very different. Particularly prominent and pervasive in the gamma band is the '40 Hz' activity. A natural hypothesis is that the Freeman oscillator lies behind this gamma activity. The figure of 63 Hz calculated for the

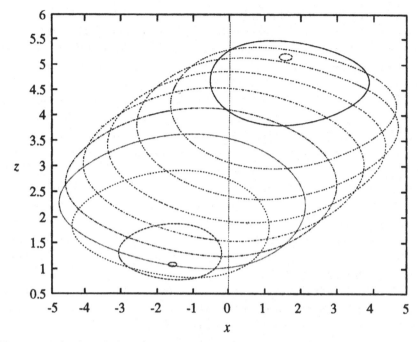

Fig. 11.6. A plot of the x/z orbits for the Freeman oscillator for the bias values $u = 0.8, 1.0, 1.5, 2.0, 2.5, 3.0, 3.5, 4.0$ and 5.44. The centre of the orbit increases in both coordinates with increasing u. The almost elliptical form is evidence of the almost sinusoidal nature of the limit cycle.

oscillator in the last section is somewhat high, but this would be lowered by association between oscillators, as we shall see. Note also that, for the rabbit olfactory system, Bressler (1987) gives the range of burst frequencies as 35–85 Hz, with the frequencies 56–63 Hz particularly prominent.

Collective oscillations in nonlinear systems are often regarded as associated with 'entrainment', i.e. with the spontaneous binding of individual somewhat-oscillatory units to a common frequency and even a common phase. This entrainment can be remarkably complete; the stock examples are the synchronisation after some time of the flashing of fireflies or of the croaking of frogs — clearly and strikingly observable. (See e.g. Chapter 9 of Beckerman (1997) and references, notably the early one to Winfree (1967) and those to Strogatz *et al.* (1988, 1989).) One may ask, then, whether oscillations in neural populations imply an entrainment of individual neurons, in that the firing of individual neurons is synchronised. This seems not to be the case in general. In the 40 Hz band individual neurons are observed to fire only about once in every five cycles, and even then not at a consistent point in the cycle (Freeman (1992, 1994)). Of course, there is a degree of entrainment, to the extent that firing is by definition more likely at the peaks of the cycle, when activity is maximal. However, this is not the rigid entrainment of the fireflies.

More generally, some workers seek special mechanisms which will explain synchronisation of neural response over a system. As far as neural masses are concerned (rather than individual neurons) the mechanism of system (14) is sufficient to bring about entrainment if the synaptic matrix W is such as to give the system sufficient connectivity. However, the PMA when implemented in neural form brings with it another and powerfully synchronising feature. This is the standardisation of global activity brought about by the standardising factor θ of relations (8.18) and (8.19) or, equivalently and more 'neurally', the adaptive biasing potential q of (8.24). We consider this mechanism in Section 12.9 and discuss other suggestions from the literature in Section 12.10.

A feature of EEG data is the appearance of 'burst frequencies': of short bursts of gamma-wave activity which end as abruptly as they began. Freeman (1992) has explained these nicely; see also Freeman and Schneider (1982). A common manifestation takes the form illustrated in the first trace of Fig. 14.6, where there is a slow periodic variation (associated simply with the animal's breathing) on which is superimposed a burst of 40 Hz variation at each peak. Consider the Freeman oscillator (15) in the case when the external input u shows slow periodic variation. When u rises through the critical threshold value oscillation will then have time to establish itself, to subside later when u sinks again through the threshold. This then explains the burst phenomenon exactly and economically. The pattern evident in the first trace of Fig. 14.6 is that which is seen at a lower level (e.g. in the olfactory bulb, in the case of the olfactory system) whereas at a higher level (e.g. in the cortex) only the bursts survive, and not the pronounced swell of the basal cycle itself.

Remarkably, the same effect is predicted from the activity standardisation of the PMA. As we shall see in Sections 12.9 and 13.4, the adaptive standardisation produces a slow global oscillation of an almost square-wave form: the *escapement oscillation*. This in its peak phase triggers bursts of oscillation, just as does the pulmonary cycle. Indeed, one is impelled to consider whether the pulmonary cycle might not be regulated by the escapement oscillation.

Such phenomena show how inadequate the analysis of a time trace in frequency (Fourier) terms can be. A Fourier analysis of the trace in Fig. 14.6 would show something of the 40 Hz variation and something of the frequency associated with respiration, but also a great many apparent beat frequencies. The actual waveform and its simple explanation would be well concealed.

Chapter 12

Freeman Oscillators — Solo and in Concert

12.1. Systems of neural units

We return to the consideration of a set of N interacting neural units, with behaviour governed by the vector equation

$$\mathcal{L}x = Wf(x) + u \tag{12.1}$$

deduced in Section 11.4. Here x is the N-vector (x_j), where x_j is the average cell current for the neurons of unit j, while $f(x)$ is the vector of $f(x_j)$ and u is the vector (u_j) of external inputs to the system. The elements of both $f(x)$ and u have the character of pulse rates and the right-hand member of (1) represents the driving effect of these pulse rates on individual units.

The element w_{jk} of the synaptic matrix W represents the strength (and sign) of the connection from unit k to unit j. If unit k is excitatory then w_{jk} will be positive and the effect is one of positive feedback (between units of the system). If unit κ is inhibitory then w_{jk} will be negative and the effect is one of negative feedback.

One could well have different activation functions for the different units, when the jth element of $f(x)$ would be written $f_j(x_j)$. For example, it has been observed (e.g. Baird (1986), Eeckman and Freeman (1991)) that inhibitory units sometimes communicate their output to neighbouring units by a direct electric effect rather than by use of the membrane/axon/synapse link. This would not be feasible over large distances, for the reasons explained in Section 11.1, but represents a structural economy over short distances.

Positive feedback produces the reinforcement required to hold an image in memory when the sensory input has ceased. It must be supplemented by negative feedback if oscillatory function is to be achieved. Negative feedback is also used to achieve the 'on-centre off-surround' response which is believed to play an essential part in the function of the retina and other organs. This is the arrangement by which feedback is positive between units which are physically close and negative between units which are physically distant. It has the effect of increasing contrast, and so of emphasising strong features and suppressing weak ones (see e.g. Grossberg (1973) and Amari (1980, 1983)). For the same reason, it is a mechanism frequently invoked for models of pattern formation and morphogenesis (see e.g. Murray (1989)).

The PMA of Chapter 8 achieves such a global inhibition, but does so by a mechanism which is multiplicative rather than additive. The adaptive standardising factor θ in the feedback loop of (8.18) and (8.19) automatically achieves both the standardisation of variables and the selection of the most active channel which the Grossberg mechanisms of Section 3.6 were designed to bring about. The effect is realised most elegantly by an adaptive global bias on potentials; see (8.24) and Section 8 of this chapter.

In Section 11.5 we considered the linearised version and stability properties of model (1) for the particular case of the Freeman oscillator. A good part of that analysis can be generalised.

Suppose that u is constant in time. If x then settles to an equilibrium value $\bar{x}$ this must satisfy the equation

$$L(0)\bar{x} = Wf(\bar{x}) + u. \tag{12.2}$$

Equation (2) may have many solutions, and these may be stable or unstable. If there are no stable equilibria at all then x must ultimately settle to a motion which at the least has finite amplitude: the stability of the operator $\mathcal{L} = L(\mathcal{D})$ plus the boundedness of f will ensure that x is effectively confined to some bounded region in $\mathbb{R}^N$.

We test for stability of $\bar{x}$ by the usual linearisation. If $x(t)$ is perturbed from $\bar{x}$ by a small amount $\xi(t)$ then this obeys the linearised equation

$$\mathcal{L}\xi(t) = WG\xi(t), \tag{12.3}$$

where G is the diagonal matrix with jth entry $\gamma_j = f'_j(\bar{x}_j)$. If $\xi(t)$ takes the form ξe^{st} for fixed ξ then (ξ, s) obey the equation

$$L(s)\xi = WG\xi. \tag{12.4}$$

The possible s-values are then the roots of any of the equations

$$L(s) = \lambda_j, \quad (j = 1, 2, \ldots, N), \tag{12.5}$$

where the λ_j are the N eigenvalues (in general complex-valued) of the matrix WG. The general solution of (3) is

$$\xi(t) = \sum_j \sum_k c_{jk} \xi_j e^{s_{jk}t},$$

at least in the case of distinct s_{jk}. Here the c_{jk} are arbitrary scalars, ξ_j is the eigenvector corresponding to eigenvalue λ_j, and the s_{jk} are the solutions of (5) for given j.

Stability of $\bar{x}$ requires that all the s_{jk} should have strictly negative real part. This can be tested in terms of the eigenvalues λ.

Theorem 12.1. *Suppose that $L(s) = s^2 + cs + d^2$, where c and d^2 are real and positive, and consider the equation in s:*

$$L(s) = \lambda = \mu + i\nu. \tag{12.6}$$

All roots of this equation have strictly negative real part if and only if

$$d^2 - \mu > (\nu/c)^2. \tag{12.7}$$

More specifically, if the set of (μ, ν) values specified by (7) is regarded as region S in the complex λ-plane, then the complementary region $\bar{S}$ is the image under the mapping $L(s)$ of the closed half-plane $\mathrm{Re}(s) \geq 0$. If equality had held in (7) then equation (6) would have had s-roots with zero real part; if the reverse inequality had held then (6) would have had roots with positive real part.

Proof. The boundary

$$d^2 - \mu = (\nu/c)^2, \tag{12.8}$$

separating S and $\bar{S}$ is indeed the image of the imaginary s-axis under the mapping L. For, suppose that $L(s) = \mu + i\nu$. If we set $s = i\omega$ for real ω then $\mu = d^2 - \omega^2$ and $v = c\omega$. Eliminating ω between these two relations we obtain equation (8), describing the locus of $\lambda = L(i\omega)$ as ω varies. This locus is illustrated in Fig. 12.1.

The full theorem will be demonstrated if we can prove that, if the equation $L(s) = \lambda$ has an s-root in the closed right half-plane $\mathrm{Re}(s) \geq 0$, then λ must lie in $\bar{S}$.

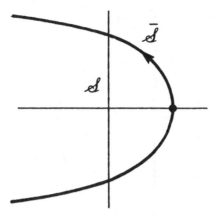

Fig. 12.1.   A sketch of the regions S and $\bar{S}$ defined in Theorem 12.1. The upper part of the boundary, drawn heavily, is the locus in the complex plane of $L(i\omega)$ as ω increases from 0 to ∞.

Suppose that $s = \chi + i\omega$. Then, in a generalisation of (8), we find that the locus of $\lambda = L(s)$ as ω varies and χ is held fixed is

$$d^2 - \mu + \chi^2 + c\chi = \left(\frac{\nu}{c + 2\chi}\right)^2. \tag{12.9}$$

For $\chi \geq 0$ the parabolas (9) all lie in $\bar{S}$, and indeed cover it simply. That is, $\bar{S}$ is indeed the image of the set $Re(s) \geq 0$, as asserted.   □

The image of the left half-plane $Re(s) < 0$ is S and part of $\bar{S}$, a reflection of the fact that for some λ the equation $L(s) = \lambda$ can have roots in both half-planes.

There is a more elegant proof, based on a view of the situation which will prove powerful in later examples. This is to note that the D-loop around the right half s-plane in Fig. 12.2 (regarded as an infinite loop, although we draw it as a finite one) maps under the transformation $\lambda = L(s)$ into the directed contour of Fig. 12.3. That this contour does not contain the λ-origin corresponds to the fact that $L(s)$ has both its zeros in the left half-plane. (The reader might like to confirm that the D-loop around the *left* half s-plane maps into a contour encircling the λ-origin twice.)

The effect of varying the driving term u in (1) depends on the character of the synaptic matrix W, but the following rough picture will often hold. If the value of u is low then the elements of $\bar{x}$ are also low, lying in a region where f saturates to the left and has small gradient. That is, the elements of G, non-negative in sign, are small. As u increases then so will $\bar{x}$ (at

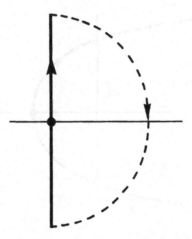

Fig. 12.2.   The 'infinite' D-loop around the closed right half of the complex s-plane.

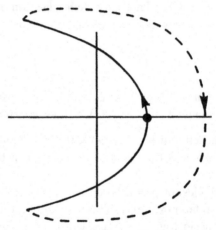

Fig. 12.3.   The contour into which the D-loop of Fig. 12.2 maps under the transformation $L(s)$.

least in the cases when (2) has a unique solution). If $\bar{x}_j$, say, continues to increase, then γ_j will increase to a maximum as $\bar{x}_j$ passes through the point of inflection of f, and will then decrease.

The effect will be that the eigenvalues λ of WG are initially small, and, if they are small enough that they lie inside $\mathcal{S}$, then the initial equilibrium $\bar{x}$ is stable. However, if W is such that the increase in G takes any of the eigenvalues of WG out of $\mathcal{S}$ then an instability will set in.

If the crossing point is complex then the instability will manifest itself in the form of a periodic oscillation, initially of small amplitude and sinusoidal character. This is just the Hopf bifurcation of Chapter 4. Once one gets into the region of real instability then, of course, the linearised model is no longer adequate. However, even in the clearly nonlinear regime, the saturating effects of the sigmoid $f(x)$ will ultimately confine x to a bounded set in $\mathbb{R}^N$. If one or more eigenvalues move into $\mathcal{S}$ then one has the alternatives of a straight instability (which usually means a switch to another equilibrium of (2)), a limit cycle, or a more complex oscillation which may be chaotic.

One could regard the operation of the net as a sequence of codings $u \to \bar{x} \to G \to$ dynamic behaviour. That is, at least in the linearised treatment, one can regard the effect of u (supposed fixed) upon dynamics as being mediated first through the location of the equilibrium $\bar{x}$ and then through the consequent values of the diagonal matrix of gradients G, which rescales the elements of the synaptic matrix W. (Note that this could be regarded as a symmetric rescaling: the eigenvalues of the matrix WG are the same as those of the matrix $G^{l/2}WG^{1/2}$.)

However, to say that the dynamic behaviour of the net varies with the value of u says nothing. If the net is to operate as an associative memory then one seeks the stronger property: that the net should classify the input u into meaningful classes, by adopting a dynamic behaviour appropriate to the *class*. The possibility of such behaviour is exhibited more clearly in the case when the data are injected through initial values $x(0)$ rather than through a driving input u. Let us annul u by setting it equal to zero, say, and suppose that the consequent version of equation (2) for $\bar{x}$ has many solutions. This will indeed typically be the case if there is positive feedback; see the next section for the simplest example. Then every such solution is potentially the centre of a basin of attraction, defining a dynamic regime. The input $x(0)$ is then classified by the basin into which it falls, which is in turn quickly revealed by the dynamics which develop.

12.2. The case of simple positive feedback

The simplest nontrivial case, simpler even than the Freeman oscillator, is that of a single unit with feedback. In this case (1) reduces to the scalar relation

$$\mathcal{L}x = wf(x) + u. \tag{12.10}$$

Suppose that the unit is an excitor, so that $w > 0$ and feedback is necessarily positive. This then is the system that Freeman would denote KI_e: a neural unit with positive feedback (achieved by excitatory connections between individual neurons of the unit). In the inhibitory case $w < 0$ the system would be denoted KI_i.

The possible equilibrium values $\bar{x}$ must be roots of

$$d^2 x = w f(x) + u, \qquad (12.11)$$

and we now find ourselves on very familiar territory. The roots of (11) correspond to the intersections of the two graphs in Fig. 12.4, and there may be as many as three. One would conjecture that the only roots yielding stable equilibria are those for which the sigmoid crosses the straight line from above, and this is exactly the criterion yielded by Theorem 12.1. There is only the single eigenvalue $\lambda = w\gamma = w f'(\bar{x})$. This is real, and criterion (7) then reduces to $d^2 - w f'(\bar{x}) > 0$, which is exactly the condition conjectured.

The real nature of λ implies that, if stability fails, then there is no oscillation; x merely slides to a stable equilibrium value.

In Section 3.4 we mentioned the role of such a device as a simple bistable system. We also observed thai, if w and u are chosen so that all three roots of (11) coincide (i.e. so that the straight line is tangent to the sigmoid at its inflection point, as in Fig. 12.5), then $\bar{x}$ will show great sensitivity to variation in u about this particular value.

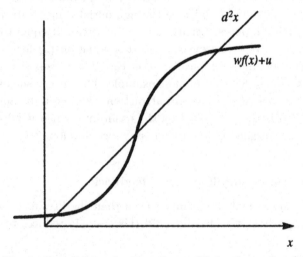

Fig. 12.4. The familiar diagram locating the equilibria of equation (11), with the familiar conclusion that only the two outer equilibria are stable.

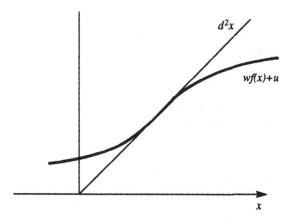

Fig. 12.5. The situation of Figure 12.4, but so arranged that the three equilibria are coincident, implying sensitive dependence of the equilibrium value $\bar{x}$ on the value of u in that neighbourhood.

12.3. Systems of coupled oscillators

We wish ultimately to set up an associative memory based on the Freeman oscillator. In seeking an appropriate structure we shall be guided both by the structures derived for the ANN case from the considerations of Chapters 7 and 8 and by the actual anatomy of organisms. Prominent in the olfactory bulb, for example (see Chapter 14), is a feature which resembles an array of Freeman oscillators, symmetrically linked via their excitatory elements. The effect of inhibitory elements seems to be short-range (in contrast to the situation for the visual system) and mediated electrically rather than axonally. We shall then consider the class of models for which the vectors x and z of the cell currents of excitors and inhibitors obey the equations

$$\mathcal{L}x = Wf(x) - z + u$$
$$\mathcal{L}z = f(x). \tag{12.12}$$

Here W is a symmetric matrix, representing synaptic connections between excitors (and so symmetric positive feedback). We have taken the extreme short-range case, in which an inhibitor is connected only to its paired excitor, and have represented the electrical (rather than axonal) mediation by setting z rather than $f(z)$ in the first relation.

It is useful to repeat the treatment of Section 1 for this more structured case. Suppose the input u to be constant in time, and suppose that equations (12) then admit an equilibrium solution $(\bar{x}, \bar{z})$, not necessarily stable. Then $\mathcal{L}x$ becomes $d^2\bar{x}$ and, eliminating $\bar{z}$ from the equilibrium equations, we deduce the equation

$$d^4\bar{x} - d^2Wf(\bar{x}) + f(\bar{x}) = u \tag{12.13}$$

for $\bar{x}$. This may have several solutions and, if data are injected via initial values $x(0)$, we indeed hope for many such solutions, some of them defining dynamic basins of attraction. However, let us for the moment consider stability for a given particular solution.

If we consider perturbations ξe^{st} and ζe^{st} from $\bar{x}$ and $\bar{z}$ then these will obey the linearised equations

$$L(s)\xi = WG\xi - \zeta$$
$$L(s)\zeta = G\xi, \tag{12.14}$$

where G is the diagonal matrix with jth diagonal element $\gamma_j = f'(\bar{x}_j)$. Setting $L(s) = \lambda$ and eliminating ζ from relations (14) we deduce the equation

$$(\lambda^2 I - \lambda WG + G)\xi = 0.$$

The eigenvalues are thus the zeros of $|\lambda^2 I - \lambda WG + G|$.

A special case of interest is that in which the model is spatially homogeneous, in that the γ_j are independent of j and equal to γ, say, so that $G = \gamma I$. In this case the $2N$ eigenvalues of the system are the roots of the N quadratic equations

$$\lambda^2 - w_k\gamma\lambda + \gamma = 0, \quad (k = 1, 2, \ldots, N), \tag{12.15}$$

where the w_k are the N eigenvalues of W. If W is symmetric then these eigenvalues are real; if W is moreover positive definite then they are positive.

12.4. The Freeman oscillator with positive feedback

We should gain experience of simple systems of type (12), and the simplest is certainly that for $N = 1$. That is, a single Freeman oscillator with an additional feedback from the excitor to itself. This is far from trivial in its behaviour, and sets the pattern for a number of more general models.

Equation (12) then holds for scalar x and z, with the synaptic matrix W set equal to a non-negative scalar w, say. Equation (13) for the value of

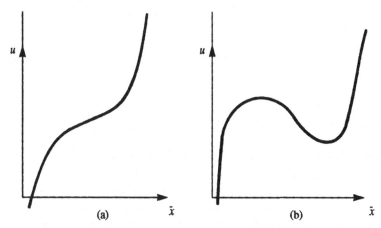

Fig. 12.6. Sketch graphs of relation (16), expressing the fixed input value u as a function of the equilibrium value $\bar{x}$ (although the causal dependence is in the reverse direction). As the feedback coefficient w increases the relation changes from the monotonic character of (a) to the non-monotonic character of (b). Inequality (17) is the condition for non-monotonicity.

$\bar{x}$ in equilibrium then reduces to

$$u = d^4\bar{x} + (1 - wd^2)f(\bar{x}), \tag{12.16}$$

a relationship which we represent in the graphs of Fig. 12.6. We see that (16) has a solution for $\bar{x}$ for any value of u, but it may under certain circumstances have three. This latter case will occur for some u if and only if

$$d^4 < (wd^2 - 1)\gamma_{\text{max}}, \tag{12.17}$$

where γ_{max} is the maximal gradient of the sigmoid f. We illustrate this case in Fig. 12.6(b).

Equation (15) for the eigenvalues λ now becomes

$$\lambda^2 - w\gamma\lambda + \gamma = 0. \tag{12.18}$$

We could investigate directly how the λ-roots of this quadratic equation lie relative to the region of stability S defined in Section 1. However, a much more illuminating and powerful method is to trace the paths of the various functions of s as s traces out the D-loop of Fig. 12.2. In particular we follow the path as s describes the imaginary axis, so that $s = i\omega$ for ω real. On this locus $\lambda = L(i\omega) = \mu + i\nu$, say, where $\mu = d^2 - \omega^2$ and $\nu = c\omega$.

By symmetry, it is enough to consider only non-negative ω. Let us write equation (18) as

$$w = \gamma^{-1}\lambda + \lambda^{-1} = g(s), \qquad (12.19)$$

say.

For definiteness, we shall take γ as an independent parameter, and investigate at what value of w and in what manner stability breaks down for a given value of γ.

The path of $\lambda = L(i\omega)$ in the complex plane as ω increases from 0 to $+\infty$ is just the upper part of the boundary of S as specified by (8), see Fig. 12.1. The path of λ^{-1} is then as in Fig. 12.7. We shall see that the path of $g(i\omega)$ for $\gamma < d^4$ is very much like that of $L(i\omega)$, but that for $\gamma > d^4$ it follows the course sketched in Fig. 12.8. For this fixed γ the system will become unstable first when the point w on the real axis in the complex

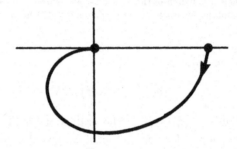

Fig. 12.7.   A sketch of the locus of $L(i\omega)^{-1}$ as ω increases from zero.

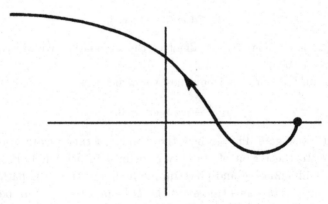

Fig. 12.8.   A sketch of the locus of $g(i\omega) = \gamma^{-1}L(i\omega) + L(i\omega)^{-1}$ as ω increases from zero. This shows the path when $\gamma > d^4$; the initial dip is absent when $\gamma < d^4$.

plane crosses the locus of $g(i\omega)$ from left to right. For the case of Fig. 12.8, we see that $g(i\omega)$ crosses the real axis at $\omega = 0$ and at just one other value of ω; let us denote this by ω_γ. There will then be instability if w exceeds the value $w_\gamma = g(\omega_\gamma)$, and the mode of instability at threshold will take the form of a sinusoidal oscillation of frequency ω_γ. One suspects that there may be a second transition when w increases through the value $g(0)$.

Theorem 12.2. *For $\gamma \leq d^4$ the critical value of w is $w_\gamma = \gamma^{-1}d^2 + d^{-2}$, and the mode of threshold instability is non-oscillatory (i.e. $\omega_\gamma = 0$). For larger values of γ the instability, when it develops, is oscillatory and of frequency equal to the sole positive root ω_γ of $|L(i\omega)|^2 = \gamma$, i.e. of*

$$(d^2 - \omega^2)^2 + c^2\omega^2 = \gamma. \tag{12.20}$$

This instability develops when w exceeds the value

$$w_\gamma = g(\omega_\gamma) = 2\gamma^{-1}(d^2 - \omega_\gamma^2). \tag{12.21}$$

The critical values w_γ and ω_γ respectively decrease and increase with increasing γ.

Proof. The path of $g(i\omega)$ crosses the real axis when $g(i\omega)$ has zero imaginary part. We have

$$\text{Im}[(g(i\omega)] = \nu[\gamma^{-1} - |L(i\omega)|^{-2}],$$

which is zero either when $\nu = 0$, i.e. $\omega = 0$, or when (20) holds. But (20) is a quadratic in ω^2 which has a positive root (and so real roots in ω) first when $\gamma > d^4$. There is only one such root and it increases with γ.

The path of $g(i\omega)$ is then indeed as sketched in Fig. 12.8, but for $\gamma \leq d^4$ the crossing point is not distinct from that at $\omega = 0$, so that $\omega_\gamma = 0$ and $w_\gamma = g(0)$.

For $\gamma > d^4$ there is a second crossing point, lying to the left of that at $\omega_\gamma = 0$, as in Fig. 12.8. The critical values are then those asserted in the theorem. We can write expression (21) for w_γ alternatively as $w_\gamma = h(\omega_\gamma)$, where

$$h(\omega) = \frac{2(d^2 - \omega^2)}{(d^2 - \omega^2)^2 + c^2\omega^2}.$$

One finds that $h'(\omega)$ has the same sign as $(d^2 - \omega^2)^2 - (cd)^2 \leq d^2(d^2 - c^2) < 0$, so that $h(\omega_\gamma)$ is decreasing in ω_γ, and so in γ. □

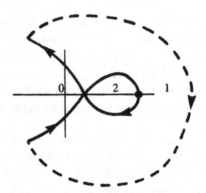

Fig. 12.9.   The locus of $g(s)$ as s traces out the D-loop of Fig. 12.2. The various regions into which this divides the complex plane have been marked with their *winding number*: the number of times the locus encircles any point in that region in a clockwise direction as s traces out the D-loop. The equilibrium under examination shows stability, non-oscillatory instability (runaway behaviour) or oscillatory instability (a limit cycle) according as the real feedback rate w lies in a region of winding number 0,1 or 2.

In Fig. 12.9 we sketch the complete g-locus as s traces out the D-loop of Fig. 12.2. For large $|s|$ the function $g(s)$ behaves as s^2, and so the infinite (dashed) part of the D-loop indeed maps into the dashed locus of Fig. 12.9. The various regions in the diagram have been marked with a number (0, 1 or 2): the *winding number*. For a point in a given region the winding number is the number of times the locus of $g(s)$ encircles that point in a clockwise direction as s traces out the D-loop. The importance of this number is immediate.

Theorem 12.3. *Suppose that w (possibly complex) lies in a region of the complex plane with winding number σ. Then the equation $g(s) = w$ has exactly σ roots with positive real part.*

Proof.   The function $w - g(s)$ is rational in s, and the winding number gives the number of zeros of this function which lie inside the D-loop less the number of poles inside the loop. But the poles of $g(s)$ are at the zeros of $L(s)$, which all lie in the left half-plane. The conclusion then follows. $\square$

If $\sigma = 0$ for the given w then the equilibrium point $\bar{x}$ is stable. If $\sigma = 1$ then there is a non-oscillatory instability, resolved by runaway to a stable equilibrium point. If $\sigma = 2$ then there is an oscillatory instability, which we expect to correspond to a limit cycle. The numerical example of the next section confirms this classification. If $\gamma < d^4$ then the inner lobe of Fig. 12.9 disappears, and there is no region of winding number 2.

If we instead take w as the independent parameter, then the conclusion is that instability will set in at an ever-decreasing value of γ as w increases, but that the frequency of oscillation, when it develops, also decreases. This oscillation is associated with a local limit cycle, one conjectures. However, when w exceeds the value $2/d^2$ then the instability is non-oscillatory, $\bar{x}$ is not the centre of any kind of basin of attraction, and the variables (x, z) change to seek such a basin.

The treatment of this section extends immediately to the special systems of oscillators considered at the end of Section 3; we simply consider the behaviour of each of the modes of the system by replacing the feedback rate w by the real eigenvalues w_k of the feedback matrix W.

12.5. Numerical results for the oscillator with feedback

Numerical test broadly confirms the analysis of the last section, but also reveals further structure. For $\mathcal{L}$ we take the standard form (11.19) and for the activation function $f(x)$ the logistic form f_1 defined in Section 11.6. The maximal value of w, beyond which instability is non-oscillatory, is then $2/d^2 = 12.5$.

A value of $w = 4.83$ was chosen; this corresponds to a threshold frequency half that of the simple oscillator, and so a period of about 31.42 milliseconds. For the given $f(x)$ it should break out when u first exceeds the value of -0.386 approximately.

Calculations show that there is indeed a stable static equilibrium which holds as u is increased up to -0.3880, and also as u is decreased down to a value 1.8085. There is a degree of hysteresis at each breakpoint, however, as there is a stable orbit (limit cycle) for u between -0.4055 and 1.8255. That is, there is a range of u, albeit narrow, for which both the static equilibrium and the limit cycle are stable. When a shift from one to the other is forced there is then a jump-discontinuity. This is a first indication that the Freeman oscillator shows its nonlinear character much more strongly when positive feedback is added.

The amplitude and period are graphed in Fig. 12.10. In contrast to the rise and fall in amplitude with increasing u predicted from a perturbation model and observed in Fig. 11.3, the amplitude is now fairly constant over the oscillatory range. This is partly, but only partly, explained by the overlap of the sets of u-values for which the static equilibrium and the cycle are stable.

We saw from Fig. 11.3 that the period of the simple Freeman oscillator varied little over the oscillatory range. In the present case it is close to the

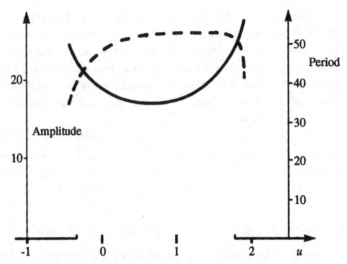

Fig. 12.10. A plot of amplitude and period for the limit cycle (of the excitor variable x) as a function of u for the case $w = 4.83$. Stable fixed points exist for $u \leq -0.388$ and for $u \geq 1.8085$. A limit cycle exists in the range $-0.4055 \leq u \leq 1.8285$. The amplitude (dashed line) of the cycle over this range is much more uniform than it was in the absence of feedback (see Fig. 11.4); it varies between 17 and 22. However, the period (full line) varies from 36 milliseconds in midrange to about 55 milliseconds at the ends of the range, before dropping very rapidly to zero.

linear prediction in the centre of the range (about 35 milliseconds) where the oscillation is also fairly sinusoidal in character. However, it increases to about 55 milliseconds at the ends of the range, where the oscillation also becomes clearly non-sinusoidal, as we see from the phase-plot for x in Fig. 12.11. This is strongly non-elliptic, and x slows down markedly as it goes through the neighbourhood of the incipient equilibrium point. Nonlinearity evidently has strong secondary effects in this feedback version.

A higher value $w = 9.58$ was tested, which should have given a threshold frequency one quarter that of the free oscillator. However, in all cases tested the operating point jumped to a solution of (16) which was stable, and so non-oscillatory.

This is in fact consistent. In Fig. 12.12 we have characterised stability properties in terms of the two basic parameters: the feedback rate w and the gradient γ of $f(x)$ at the operating point. The plane breaks into three regions $\mathcal{S}_\sigma$, corresponding to winding numbers $\sigma = 0, 1$ and 2. The boundary B_1 of the set with winding number 1 is determined by $w = g(0)$, or

$$w = d^2\gamma^{-1} + d^{-2}. \tag{12.22}$$

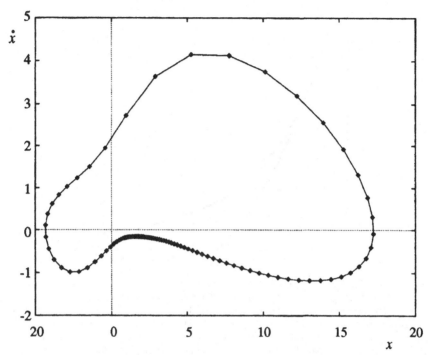

Fig. 12.11. A phase diagram for the Freeman oscillator with feedback. For a range of values of u the plot of x versus $\dot{x}$ is almost elliptical, indicating that the variation of x in the limit cycle is almost sinusoidal. However, in this diagram, for $w = 4.83$, u has the value 1.825, near the upper limit of stability for the cycle. The plot is strongly non-elliptic, and shows a tendency for x to linger at the incipient stability point.

The boundary B_2 which separates regions of winding numbers 0 and 2 (and on which the Hopf bifurcations occur) is determined by $w = g(i\omega_\gamma)$, where ω_γ is the first positive value of ω for which $g(i\omega)$ is real. These boundaries lie as illustrated. Of course, not all (w, γ) values are possible; there will be a restriction $\gamma \le \gamma_{\max}$, where $\gamma_{\max}$ is the maximal gradient of $f(x)$.

Recall that we associate winding numbers 0, 1 and 2 with stability, runaway instability and a limit cycle respectively. This is largely borne out by the calculations. All the points tested for $w = 4.83$ lay in $\mathcal{S}_0$ for low u, then in $\mathcal{S}_2$ as u was increased, then in $\mathcal{S}_0$ again as u was increased yet further. Dynamic behaviour was consistent. None of the points tested for $w = 9.58$ happened to lie in the narrow part of $\mathcal{S}_2$ which was available; either they lay in $\mathcal{S}_0$ or lay initially in $\mathcal{S}_1$ and then jumped promptly to an operating point in $\mathcal{S}_0$.

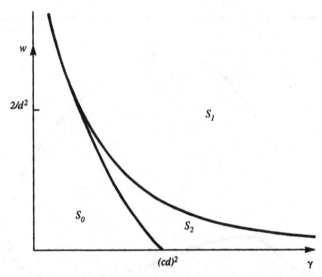

Fig. 12.12. The decomposition of the (w, γ) plane into sets S_σ corresponding to the winding number σ. (That is, the number of times $g(s)$ encircles the value w for the given value of γ as s describes the D-loop of Fig. 12.2.) For $\sigma = 0, 1$ and 2 these are predicted as the sets of (w, γ)-values for which the equilibrium under examination shows stability, runaway instability or a limit cycle. This prediction is confirmed by the numerical solutions, apart from a degree of ambiguity at the boundaries.

However, the hysteresis effect indicates that this picture holds only as a first approximation. The regions S_0 and S_2 do not overlap, but the regions for which the static equilibrium and the cycle are stable do in fact show a small overlap. This is a matter for further investigation.

12.6. A block of oscillators

A step up in complication is to consider the N-oscillator system with $w_{jk} = w/N$ for all j and k, so that $W = wE/N$, where E is the $N \times N$ matrix of units. That is, the feed from oscillator j to oscillator k has a common synaptic strength w/N for all j, k. We shall term such a system a *block* of oscillators for conciseness. The interest is to see whether some collective activity prevails over the individual activity in such a case.

If the elements u_j of the external input u are constant and independent of j then one potential equilibrium of interest is certainly that for which the static equilibrium values $\bar{x}_j$ and $\bar{z}_j$ are also independent of j, with the implication that $G = \gamma I$, and the eigenvalues λ are the roots of equations (15).

The matrix E has a single unit eigenvalue (with a vector of units as eigenvector) and $N - 1$ zero eigenvalues (with eigenvectors whose elements sum to zero). It follows then that one pair of eigenvalues are just the roots of equation (18) and the remaining $2N - 2$ are the roots of $\lambda^2 + \gamma = 0$, counted $(N - 1)$-fold. The first pair of eigenvalues correspond to a collective mode of vibration in the system, and this mode has the same dynamics as the model of the last section: a single Freeman oscillator with positive feedback. Remaining eigenvalues correspond to deviation of the motion of individual oscillators from this collective motion, and are equivalent to the variation of $N - 1$ independent standard Freeman oscillators.

Since the threshold value of the gradient γ decreases with increasing w we see that it is the collective motion which is triggered most easily, and that this is at a lower frequency than that of a free oscillator.

Calculations were made on this with the parameters of Section 8 and with $w = 4.83$ and $N = 8$. The eight oscillators showed perfect synchrony, so the effect was one of a single oscillator with feedback.

12.7. A chain of oscillators

The block model of Section 6 is a very unstructured one, in that connection strengths (and delays) are assumed not to depend at all upon the spatial disposition of the units. The simplest model which gives some spatial structure is that in which the oscillators are assumed to form a chain, with connections only between nearest neighbours. In fact, the analysis is somewhat simplified if we assume that the ends of the chain are joined, so that the TV oscillators form a ring. This 'periodic boundary condition' gives a homogeneous spatial structure. The first relation of (12) then becomes

$$\mathcal{L}x_j = -z_j + \frac{1}{2}w[f(x_{j-1}) + f(x_{j+1})] + u_j, \quad (j = 0, 1, \ldots, N - 1),$$

when written out element by element. The periodic boundary condition implies that x_N is to be identified with x_0, etc.

If we suppose spatial homogeneity in equilibrium then we shall have $G = \gamma I$ again. The elements w_{jk} of W are $w/2$ if $j - k = \pm 1 \pmod{N}$, zero otherwise. One confirms easily that, if $\chi = \exp(2\pi ji/N)$, then W admits an eigenvector with jth element χ^{jk}, and that the corresponding eigenvalue is

$$w_k = w\cos(2\pi k/N). \tag{12.23}$$

We obtain the full set of eigenvectors by choosing $k = 1, 2, \ldots, N$.

The eigenvalues are thus the roots of the equations (15) with the w_k having the evaluations (23). The characterisation of their behaviour and its implications again follow from Theorem 12.2. The mode which is actuated first as stimulation increases is that for which w_k is largest. This is w_N, if we assume w positive; we can consistently write it as w_0. The mode $k = 0$ is again the collective mode, in which all oscillators beat together at a common phase (and at a frequency lower than that of the free oscillator).

However, the modes become ever more densely packed as N increases (in that their critical thresholds become ever closer), so one can expect that, for large N, all the small-k modes (i.e. those whose amplitude and phase vary only slowly in space) will be present in the motion to some degree.

12.8. Feedback plus standardisation: An essential mechanism

The feedback effects considered hitherto have been purely additive, and inhibition has been confined to the local inhibition needed for the Freeman oscillators. However, we know that the PMA of Chapter 8 applies a global multiplicative inhibition, in that feedback is standardised by the presence of the factor θ in equations such as (8.18) and (8.19), or by a general depression q of the potential x, as in equation (8.24). We shall examine a mechanism of this type, in preparation for the application to associative memories in the next chapter.

We shall consider a neural/oscillatory implementation of the PMA in the next chapter. Suffice it to say for the moment that a simplified version of relations (8.24) would be

$$\mathcal{L}x = d^2 w f(x - q), \quad \mathcal{D}q = \kappa[f(x - q) - 1] \qquad (12.24)$$

These represent a single neural unit with a feedback term wf, but with its operating point standardised adaptively by a bias term q. In fact, the second relation varies q in an attempt to bring $x - q$ to the value δ for which $f = 1$. We may suppose (with appropriate scaling) that δ is the chosen operating point, at which f has gradient γ. System (24) has an equilibrium at $x = w, q = w - \delta$.

The obvious oscillatory version of system (24) would be

$$\begin{aligned}
\mathcal{L}x &= d^2 w f(x - q) - z \\
\mathcal{L}z &= f(x - q) \\
\mathcal{D}q &= \kappa[f(x - q) - 1].
\end{aligned} \qquad (12.25)$$

This then appears as a Freeman oscillator with positive feedback, but also with the standardising bias q which aims to bring the operating point $x - q$ to δ. The system has its only equilibrium at $x = q + \delta = w - d^{-4}$, $z = d^{-2}$, $z = d^2$. The value of the feedback coefficient w may be quite large, in that it is proportional to n, the dimension of the data vector. If there were no standardisation then this would lead to a runaway instability (as distinct from the oscillatory instability which is the prelude to a limit cycle) as it would bring w into the region of winding number 1 of Fig. 12.9. However, there is now the possibility that standardisation could remedy this.

It will turn out in fact that adaptive standardisation with any positive correction rate κ removes the possibility of a runaway instability altogether. However, if κ is chosen too large (i.e. if correction is too rapid) then q tracks x so well that the operating point is held close to δ and the system cannot oscillate. These are the aspects we wish to study.

Consider perturbations ξe^{st}, ζe^{st} and χe^{st} of x, z and q from their equilibrium values. Then these obey the equations

$$\lambda \xi = (d^2 w \gamma)(\xi - \chi) - \zeta$$
$$\lambda \zeta = \gamma(\xi - \chi)$$
$$s\chi = \kappa(\xi - \chi),$$

where we have again written $L(s)$ simply as λ. Eliminating ξ, ζ and χ we obtain the s-determining, equation

$$\kappa \gamma \lambda^2 + s[\lambda^2 - (d^2 w \gamma)\lambda + \gamma] = 0. \tag{12.26}$$

This is rather a nice modification of the equation (18) for the unstandard-ised oscillator. We can write it in the form

$$d^2 w = \gamma^{-1}\lambda + \lambda^{-1} + \kappa \lambda s^{-1} = g(s) + \kappa s^{-1} L(s) = h(s),$$

say, where $g(s)$ is the function defined in (19). Suppose that $\kappa > 0$. Then examination shows that, as s describes the indented D-loop of Fig. 12.13, $h(s)$ traces out the locus of Fig. 12.14, with winding numbers marked. These should be regarded as modifications of Figs. 12.2 and 12.9 respectively. The indentation corresponds to a small side-displacement of the imaginary s-axis, which one could regard as equivalent to introduction of a stabilising term $-\kappa' q$ in the right-hand member of the last equation of (25). This displaces the singularity at $s = 0$ into the left half-plane, at $s = -\kappa'$, but in the end one can safely let κ' tend to zero.

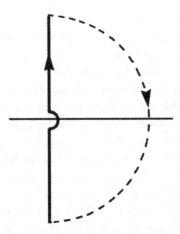

Fig. 12.13.   The indented D-loop in the complex s-plane; a modification of Fig. 12.2.

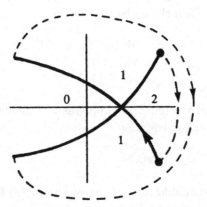

Fig. 12.14.   The radical modification of Fig. 12.9 effected by the introduction of standardisation in the oscillator/feedback model. The region of winding number 2 (for which there is a limit cycle) has now swollen to fill much of the right half-plane, so that there is no real w for which the winding number is 1, and so for which there is runaway instability.

The effect of standardisation on the graph of Fig. 12.9 is more radical. The crossing of the real axis at $\omega = 0$ is now blown out into an excursion of the locus to the points $g(0) + \kappa(c \pm i\infty)$, closed by an infinite half-loop which is the image of the indenting half-loop in Fig. 12.13. The effect is that the region of winding number 1 has been pushed out to infinity along the real axis. That is, the adaptive standardisation proofs the system against a runaway instability for any finite real w and for any positive κ.

However, one will want w to be in the region of winding number 2 (under the mapping $h(s)$) if the system is to oscillate. That is, one must have

$$d^2 w > h(i\omega_1) = g(\omega_1) + \kappa c + i\kappa[\omega_1 - (d^2/\omega_1)], \qquad (12.27)$$

where ω_1 is the (unique finite) value which makes the right-hand member of (27) real. This sets an upper bound on κ for given w and γ (or a lower bound on w for given κ and γ) if there is to be oscillation. Note that the effect of increasing κ is to bring ω_1 nearer to d, the threshold frequency of the free oscillator.

We have left verification of the sketch of Fig. 12.14 to the reader. We note merely that it is easily verified that $\text{Im}[h(i\omega)]$ is increasing in ω for ω positive, and has opposite signs for ω small and large. There is then just one crossing point of the real axis.

12.9. Numerical results for the standardised oscillator

The slightly modified version

$$\begin{aligned}
\mathcal{L}x &= wf(x - q) - z \\
\mathcal{L}z &= f(x - q) \\
\mathcal{D}q &= \kappa[f(x - q) - 0.5]
\end{aligned} \qquad (12.28)$$

of system (25) was solved numerically for the standard choices of $\mathcal{L}$ and $f(x)$ adopted in Section 11.6. The feedback coefficient w was given the value 15, which would take the unstandardised oscillator well into the region of runaway instability, and $f(x - q)$ was given the target value of 0.5 to bring the operating point to $\delta = 0$, where $f(\delta)$ has maximal gradient. The solution was calculated for a range of correction rates κ.

As predicted in the last section, there was no runaway instability for any positive value of κ. For κ exceeding a value of about 16.2 the static equilibrium was stable; a limit cycle appeared for values of κ below this value. The variation of period with varying κ is graphed in Fig. 12.15. At $\kappa \approx 16.2$ oscillation breaks out at the frequency of the free oscillator: 63 Hz, or a period of 15.7 milliseconds. This period lengthens as κ decreases, following very much the course predicted by the determination of ω_1 in (27) and tending to infinity as κ approaches zero.

However, it is the form of the oscillation itself which is interesting. In Fig. 12.16 the waveform of x is shown for $\kappa = 1$, when the oscillation has frequency of about 5 Hz and so a period of 200 milliseconds. The oscillation

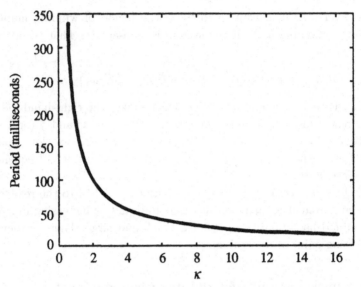

Fig. 12.15.   The variation of period of the 'escapement oscillation' with κ, the correction rate, for $w = 15$. For κ large enough, there is no oscillation. As κ decreases through a threshold value of about 16.2 an oscillation of the period of the free oscillator appears (15.7 ms, or a frequency of 63 Hz). This period lengthens indefinitely as κ approaches zero, but a limit cycle continues to exist for all positive κ.

is very nearly a square wave; it swings between two quasi-fixed points for x and differs from a square wave principally by an overshoot of about 5% at each transition. That something so 'biological' in its derivation and of no more than fifth order in its dynamics should produce a waveform so nearly square is remarkable.

One can see that such an oscillation, global in character and sharp in its transitions, would have a strongly synchronising effect in a larger system. We shall presume that it has such a function by terming it the 'escapement oscillation', generated by the feedback loop of the adaptive standardisation. That it should be so near in frequency (in the theta range, in this example) to the breathing rhythm whose effect Freeman observed in the olfactory bulb is a coincidence: the value $\kappa = 1$ was chosen simply because it fell conveniently in the test range. Nevertheless, it is hard to believe that there are two such slow modulators — perhaps one entrains the other. For the associative memory models which we shall consider in the next chapter, the escapement oscillation will be carried by the oscillators of the most active channel; see Section 13.2.

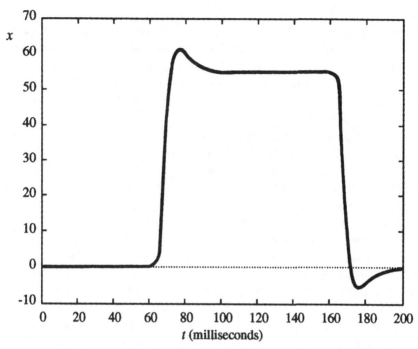

Fig. 12.16. The form of the oscillation in x for the standardised oscillator with $w = 15$, $\kappa = 1$ (the 'escapement oscillation'). This has a period of about 200 milliseconds (i.e. a frequency of about 5 Hz). The waveform itself is the remarkable feature: almost perfectly square, as x switches between two quasi-fixed points.

12.10. Some notes on the literature

There is a very large literature on coupled oscillators. In particular, there have been perhaps hundreds of papers on systems of neural oscillators since the work of Cowan and Wilson appeared in 1972. These were initially directed to confirming the possibility of multiple equilibria of both static and periodic form, and to considering transitions between these with the associated hysteresis effects as parameters were changed (see e.g. Ermentrout and Cowan (1979)). A subsequent motivation has been the attempt to explain the all-pervasive gamma waves of the 25–90 Hz band. Another has been the attempt to explain the synchrony between waves in different parts of the cortex (say), which is observed despite the presence of substantial conduction delays.

Much of this work has been directed towards study of the visual system, which is certainly far more complex than the olfactory system we have principally in mind. In the visual case the systems of oscillators can be said to have a spatial structure, in that they constitute some kind of mapping of the visual field. One has to explain the phenomena of synchronisation and what is inelegantly termed 'desynchronisation': that seemingly the oscillators concerned with responding to a given single feature ('patches') are in mutual synchrony, but are out of synchrony with those responding to another feature.

A question has been whether these mixed effects of synchrony and its absence can be achieved by purely local connections, or whether global connections are needed. A further possible mechanism, proposed by von der Malsburg and co-authors (1981, 1986, 1992), is that of allowing the strengths of the synaptic connections to vary in response to system history, on a slower time scale than that of system dynamics but on a much faster one than that of learning. In this way the system operates at two time scales.

The paper by Wang and Ternan (1995) uses local coupling to extend and synchronise existing oscillation and a global inhibitor to deter oscillation where it has not yet begun, so achieving asynchrony between patches. The paper by Wang (1995) uses adaptive couplings rather than a global inhibitor to achieve the same effect — essentially a fast Hebbian learning. Campbell and Wang (1996) use both mechanisms in the one system. Konig and Schillen (1991) employ oscillators with delay, and try to achieve patchwise synchrony by matching this delay to conduction delay between neurons. Noonburg (1997) demonstrates that a great variety of periodic effects can be achieved if adaptive connections are allowed.

Most of these papers have a basic oscillator model of the type

$$\dot{x} = -a_e x - f_i(z), \quad \dot{z} = -a_i z + f_e(x), \qquad (12.29)$$

with analogous relations for coupled oscillators (and with delays of the f-arguments for the König/Schillen paper). The Freeman analogue of (29) is (11.15), involving the basic second-order operator $\mathcal{L}$. Here x and z represent potentials for neural masses, i.e. for *populations* of excitatory and inhibitory neurons respectively. However, the papers by Wehr and Laurent (1996) and Traub *et al.* (1996) construct detailed models for *individual* neurons and study coupled versions of these model neurons computationally. The events they see are then the spikes of individual firings. The effect against which the neurons must now strive if they are to achieve synchrony is not randomness of operation, but conduction delay.

Wehr and Laurent see a fine structure in the firing sequences (now deterministic) which they suggest could encode features of the data.

They make the interesting observation that the structure corresponding to mixtures of stimuli seems not to be simply related to those observed for the individual stimuli.

Traub *et al.* refer to earlier papers by themselves and colleagues in which they claim to have explained gamma waves on the basis of interneurons alone. (Interneurons are essentially the granule cells, inhibitory cells which frequently possess no axon and exert their effect by short-range electrical mediation.) In the present model, which includes also excitatory cells, they find that the interneurons show a double spike at each firing, and claim that this provides a mechanism for synchrony by matching the time span of the doublet with conduction delays.

The Freeman oscillator of course gives an immediate explanation of gamma waves and their ubiquity. One may say that it does so precisely because its parameters (the coefficients of $L(s)$, as specified in (11.19)) have been matched to the physiologically observed parameters. However, the.matter lies deeper than that. For the reasons explained in Section 11.5, a second-order operator $\mathcal{L} = L(\mathcal{D})$ is necessary to achieve the special properties of the Freeman oscillator, and second-order behaviour is what is actually observed. These special properties include, for example, the fact that oscillation will break out once external bias exceeds a certain threshold, but that the frequency of oscillation is largely independent of that bias. It is interesting to note in this connection that Noonburg, using essentially a first-order operator in the place of $\mathcal{L}$, as in (29), could achieve oscillation only by introducing other mechanisms.

The simulations of assemblies of individual neurons are of course interesting; the question is whether the detail-sensitive conclusions are robust to statistical variation in neuron behaviour.

The goal driving our analysis has been rather different from that motivating most of the authors quoted above. They were looking for a mechanism, any mechanism, which would explain the synchrony/asynchrony effect. We were looking for a principle upon which to base an associative memory which would cope with fading data. The PMA principle adopted generated of itself the features of largely-local connection (as embodied in the synaptic matrix $W = A^T A$ for a reduced space of dimension p) and global modulation (as embodied in the standardising operation for which the principle calls). Synchrony was not specifically sought, but will be encouraged by the nonlinear couplings and assured by the standardising mechanism and the escapement oscillation which it generates. An asynchrony effect would have been produced as well if we had assumed that the data fell into blocks for which the corresponding components of trace were statistically independent. However, such an assumption would represent only a partial step towards something which we have explicitly forsworn for the moment: treatment of visual data, with its implications of spatial structure and non-additive features.

Chapter 13

Associative Memories Incorporating
the Freeman Oscillator

The goal of this part in general, and of this chapter in particular, is to arrive at well-founded associative memories which incorporate the Freeman principles. That is: that classes of data input are represented and internal communication is achieved by oscillatory modes of the system, and that threshold behaviour is marked by the onset of such oscillation. We take the Freeman oscillator as a principal building block of such systems, but we need guidance for the assembly of these blocks.

As we have already observed in Section 11.3, guidance can be sought in at least two ways. One is to seek methods of inference and information-storage whose performance is close to optimal for some specific models of data-generation, and then to see whether such methods have a natural network realisation. This was the course followed for the ANN context in Chapters 6–8.

The other course is to look to Nature: to examine the anatomy of animal neural systems. Freeman has himself taken the animal olfactory system as the prototype of a biological associative memory, and, for reasons to be explained in Chapter 14, this is a good starting point. However, animal anatomy is complex and subtle beyond measure, and the reasons why things are the way they are (or might be the way they seem to be) are very far from evident. That is why biology is such a huge, basic and challenging subject.

In this chapter we shall transfer the designs developed for the ANN case in Chapters 7 and 8; in the next we shall make comparisons with actual anatomy.

13.1. Safe data: Adaptation of the Bayesian inferential net

Consider first the case treated in Chapter 7: the case in which data is 'safe' in that it is preserved in its original form and can be recalled at any time in the calculations. A combination of Bayesian and robustness considerations led us in Sections 7.2 and 7.3 to the estimating relations

$$r = U A^T f(x), \quad x = y - W f(x) \qquad (13.1)$$

for the loading vector r. Here $U = V_{rr}$ and $W = A U A^T$, and the 'estimated residual' $\hat{\epsilon} = y - Ar$ has now been denoted x, for uniformity with the neural interpretation it will shortly receive.

It is the second relation of (1) which interests us: it expresses a type of Hopfield net, determining the field of 'residuals' in terms of the observational input y and inviting neural interpretation.

As proposed in (7.13), a dynamic neural version of this relation would be

$$\mathcal{L}x = d^2[y - W f(x)]. \qquad (13.2)$$

This system is stable and converges to a unique equilibrium, as demonstrated in Theorem 7.1. The matrix W is positive definite. In fact, we expect its elements to be individually non-negative, since the impossibility of negative odours will imply this property for A. The neural units of system (2) are thus mutually inhibitory, despite receiving an external input which is excitatory. Feedback is negative because the question is one of apportioning any improbability in the data beween improbability in the value of r and improbability in the value of ϵ.

A system of Freeman oscillators which is statically equivalent to (2) is

$$\mathcal{L}x = d^2(y - z)$$
$$\mathcal{L}z = d^2 W f(x). \qquad (13.3)$$

Here we have assumed, again more for simplicity than necessity, that the local action of an inhibitor is proportional rather than sigmoid (i.e. electrical rather than axonal), and that the term $W f(x)$ conveys all the effect of excitors on inhibitors which is necessary for oscillation.

The models of the last section usually supposed symmetric excitor/excitor connections, as is often asserted to be the case anatomically. Relation (3) differs in that it shows symmetric excitor/inhibitor connections.

This modification in pattern produces the effect that all excited modes have a threshold frequency equal to that of the isolated Freeman oscillator: $\omega = d$. This follows simply from the fact that the perturbation equations

(12.14) now become

$$\lambda \xi = -d^2 \zeta, \quad \lambda \zeta = d^2 WG\xi,$$

so that the eigenvalues $\lambda = L(s)$ are the roots of $|\lambda^2 + d^4 WG| = 0$. The eigenvalues of WG are the same as those of the positive definite matrix $G^{1/2} WG^{1/2}$, and are consequently positive. The eigenvalues λ are thus purely imaginary, and must yield $\omega = d$ at threshold, as in our initial treatment of Section 11.5.

This is a case that had to be followed up, and the conclusions are interesting. However, the assumption of safe data excludes what we now believe to be the essential challenge for an associative memory.

13.2. The oscillatory version of the one-stage PMA

If an associative memory is to overcome the problem of fading data then it must incorporate positive feedback, in order to retain the essential features of the data while the inference is being formed. The probability-maximising principle proposed in Chapter 8 automatically led to such a structure in the algorithm (the PMA) which it generated. We regard the PMA as a solution of the problem which is both natural and (by the analysis of Chapter 9) asymptotically optimal, and so now seek an oscillatory version of it.

Let us first consider the case of deciding between p simple traces a_j, the columns of A. We term this the 'one-stage' case in that it calls for only one bank of neurons. The PMA equation for the idealised data vector y then simplifies to (8.16):

$$\dot{y} \propto \theta A \phi f A^T y - y. \tag{13.4}$$

Here f is the familiar sigmoid activation function, acting on the vector following it, and ϕ is the diagonal matrix with elements $\phi_j = \pi_j \exp(-\frac{1}{2}|a_j|^2/v)$. The quantity π_j is the prior probability of trace a_j, and θ is a normalising factor $(\mathbf{1}^T \phi f A^T y)^{-1}$.

In terms of the reduced variable $x = A^T y$ (4) becomes

$$\dot{x} \propto \theta W \phi f x - x, \tag{13.5}$$

where now $W = A^T A$. As we saw in Section 8.4, this relation amounts to a version of the Hamming net, seeking the best-matching trace by its dynamics, and somewhat blurred by the uncertainty of discrimination in the presence of observation noise. If the signal/noise ratios $|a_j|^2/v$ are large then we expect the equilibrium value of the p-vector $\theta\phi fx$ to indicate which hypothesis j has been accepted, in that it will approximately have a unit in the corresponding place and zeros elsewhere.

In Sections 8.5 and 8.6 we determined the scaling factor by a separate dynamic equation, and proposed in fact that this scaling effect should be achieved by biasing the potential x. That is, $\theta f(x_j)$ was replaced by $f(x_j - q)$, where q is varied to achieve the normalisation. We can well take this further, and replace $\theta\phi_j f(x_j)$ by $f(x_j - b_j - q)$, where b_j is a fixed bias specific to node j (equal to $-v\log\phi_j = \frac{1}{2}|a_j|^2 - v\log\pi_j = \frac{1}{2}w_{jj} - v\log\pi_j$). The neural version of (5) proposed in Section 8.6 then becomes

$$\mathcal{L}x = d^2 Wf(x - b - q\mathbf{1}), \tag{13.6}$$

where the scalar q is the global adapting bias and b is the vector of fixed local biases. The bias q will adapt to its standardising value by the feedback relation

$$\mathcal{D}q = \kappa[\mathbf{1}^T f(x - b - q\mathbf{1}) - 1]. \tag{13.7}$$

Relations (6) and (7) give a neural realisation of the PMA in this case, 'one-stage' in that there is only one bank of oscillators, and with the positive feedback matrix proportional to W.

As indicated above, one expects that, if the signal/noise ratio is high in the input data, then relations (6) and (7) will have p stable solutions, the jth corresponding to acceptance of the hypothesis that the true trace is a_j and signalled by the fact that the vector $f(x - b - q\mathbf{1})$ has approximately a unit in the jth place and zeros elsewhere. If the function f were indeed exponential then this would follow simply from the fact that the PMA has its stable solutions at the values y which maximise $\rho(y)$ locally, and that these will be approximately the basic traces, or the columns of A, if the signal/noise ratio is large. Correspondingly, the stable equilibria of x will be approximately the columns of $A^T A = W$. However, we should convince ourselves that this remains true for more general f. We state the assertion as a theorem, more to pin down the statement and its justification than to supply the tedious details of a full proof.

Theorem 13.1. *Suppose that the elements of W are of order n, where n is large, and that f is a monotone increasing function whose value tends to zero with decreasing argument Then equilibria of system (6), (7) exist in which any one element of the p-vector $\bar{f} = f(x - b - q\mathbf{1})$ is approximately unity, the others approximately zero, and x is proportional to the corresponding column of W.*

Proof.   Suppose for definiteness that it is the first element of $\bar{f}$ which is dominating. Then, by (6), x is equal to the first column of w_1 of W. We see then that the argument of $\bar{f}_1$ is greater than that of $\bar{f}_k$ by an amount $w_{11} - \frac{1}{2}w_{11} - w_{1k} + \frac{1}{2}w_{kk} + O(1) = \frac{1}{2}|a_1 - a_k|^2 + O(1)$, which is positive and of order n. This implies that $f_k \ll f_1$ for $k \neq 1$. The normalisation of $\bar{f}$ then implies that it has the form first postulated.   □

A natural oscillatory version of (6) would be

$$\mathcal{L}x = d^2\,Wf(x - b - q\mathbf{1}) - z, \quad \mathcal{L}z = f(x - q\mathbf{1}), \tag{13.8}$$

with q again determined dynamically by (7). Let us examine this for its threshold properties, the point being that we wish the homing of the system on to the equilibrium associated with a particular trace to be signalled by appearance of an oscillation pattern characteristic of that trace.

If ξe^{st}, ζe^{st} and χ^{est} are the perturbations of x, z and q from their static equilibrium values then the equations

$$\begin{aligned}
\lambda\xi &= WG\Delta - \zeta \\
\lambda\zeta &= G\Delta \\
s\chi &= \kappa\mathbf{1}^T G\Delta
\end{aligned} \tag{13.9}$$

hold, where $\Delta = \xi - \chi\mathbf{1}$ and, as ever, $\lambda = L(s)$ and G is the diagonal matrix of gradients $f(\bar{x}_j - b_j - \bar{q})$. From these we deduce the eigenvalue equation

$$|\kappa\lambda^2 EG + s(\lambda^2 I - \lambda WG + G)| = 0, \tag{13.10}$$

where $E = \mathbf{1}\mathbf{1}^T$ is the matrix of units.

As we saw in Section 12.8, we cannot neglect the dynamic nature of q by neglecting the term in κ in (10). Nor would we wish to, as it turned out to play an essential stabilising and synchronising role. However, if we suppose the equilibrium is that at which $x \approx w_1$, then it is reasonable in a first approach to assume W so large that γ_k is negligible in comparison with γ_1 ($k \neq 1$). In this case (10) reduces to

$$(\lambda^2 s)^{p-1}[\kappa\gamma_1\lambda^2 + s(\lambda^2 I - w_{11}\gamma_1\lambda + \gamma_1)] = 0. \tag{13.11}$$

The square-bracketed term is familiar to us from (12.26); it just reflects the dynamics of the oscillator associated with the contingency $j = 1$, standardised as was the oscillator of Section 12.8. The factor $\lambda^{2(p-1)}$ reflects the fact that the other $p - 1$ oscillators are inactive. Note that $w_{11} = |a_1|^2$ and that q will adjust so that the operating point for the first oscillator will be close to the desired level δ. That is, γ_1 will have the standard value $\gamma = f'(\delta)$.

The analysis of Section 12.8 applies to the square-bracketed term, and we know from that section what behaviour to expect. With parameters correctly adjusted there will be an escapement oscillation, of frequency well below the 63 Hz of the isolated Freeman oscillator. It characterises the equilibrium in that the output from the neural block (which is the output from the system; see Figure 8.4 with $R = I$) is proportional to $G\xi$. This, because of the assumed form of G, is a p-vector which is approximately zero in all places but the first. That is, even the oscillatory component of output indicates that the system has decided for the first trace.

13.3. The oscillatory version of the two-stage PMA

We saw in Sections 8.5 and 8.6 that compound traces were represented in the model simply by the substitution of AR for A. Here R is a matrix whose kth column gives the weights in which the basic traces a_j ($j = 1, 2, \ldots, p$) are to be combined in the kth admissible compound trace ($k = 1, 2, \ldots, m$). We saw that this factorisation implied a separation of the feedback loop into two stages, and that it was natural to associate a set of neural elements with each of these stages.

We consider the model thus deduced the best candidate yet for a model of the associative memory implicit in something like the olfactory system. (Models for the visual system must address the much more challenging issues of spatial relation and non-additive superposition.) In this section we shall consider an oscillatory version of this two-stage net. In Fig. 13.1 we have set out the net in block form. However, for simplicity we have not included the 'hold' unit of the Chapter 8 versions, which is primarily for purposes of data injection, and belongs to the loop more statically than dynamically.

Consider now an oscillatory version of that model. Let x_1 and x_2 be the vectors of potentials evoked by the inputs associated with the upper and lower neural arrays; these are of dimensions m and p respectively. Let z_1 and z_2 be the corresponding potentials for the inhibitory units now to be associated with the excitatory units. The full set of dynamic relations is then

$$
\begin{aligned}
\mathcal{L}x_1 &= d^2 R^T W f(x_2) - z_1 \\
\mathcal{L}z_1 &= f(x_1 - b - q\mathbf{1}) \\
\mathcal{L}x_2 &= d^2 R f(x_1 - b - q\mathbf{1}) - z_2 \\
\mathcal{L}z_2 &= f(x_2) \\
\mathcal{D}q &= \kappa[\mathbf{1}^T f(x_1 - b - q\mathbf{1}) - 1].
\end{aligned}
\tag{13.12}
$$

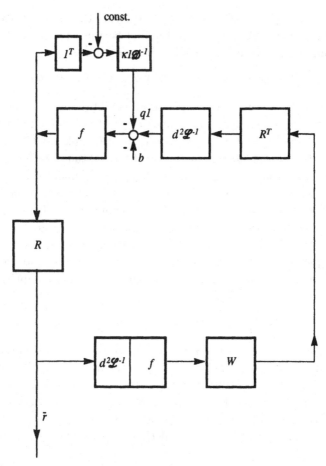

Fig. 13.1.   A block diagram of the associative memory specified by equations (8.24): the two-stage PMA with an adaptive standardisation. Equations (12) express the oscillatory version of this model.

Perturbations $\xi_k e^{st}, \zeta_k e^{st}$ and χe^{st} from the equilibrium values of x_k, y_k $(k = 1, 2)$ and q then satisfy

$$\lambda \xi_1 = d^2 R^T W G_2 \xi_2 - \zeta_1$$
$$\lambda \zeta_1 = G_1 \Delta$$
$$\lambda \xi_2 = d^2 R G_1 \Delta - \zeta_2$$
$$\lambda \zeta_2 = G_2 \xi_2$$
$$s \chi = \kappa 1^T G_1 \Delta,$$

where $\Delta = \xi_1 - \chi 1, \lambda = L(s)$ and G_1 and G_2 are the diagonal matrices of f-derivatives at equilibrium for the top and bottom arrays. We find, with some elimination of variables, that $(s\lambda^2 I + HG_1)\Delta = 0$, where

$$H = \kappa\lambda^2 E + s[I - \lambda^2 d^4 R^T W G_2(\lambda^2 I + G_2)^{-1} R].$$

Suppose, as in the last section, that equilibrium has been reached at a point corresponding to acceptance of the first compound trace (for definiteness), so that all elements of G_1 are negligible relative to the 11 element, which we shall denote by γ. The amplitude of the oscillatory component of output of the system is $RG_1\xi_1$ which, under the assumption on G_1, is proportional to the first column R_1 of R. This is just the vector of loadings for the first of the m possible compound traces. That is, the fact that the system has decided for this trace is signalled in the most directly interpretable fashion in the pattern of oscillation amplitude in the output.

The spatial nature of the oscillation is then determined; we should like to determine its temporal nature. Under the reduction assumption, on G_1 we have

$$|s\lambda^2 I + HG_1| = (s\lambda^2)^{m-1}(s\lambda^2 + \gamma H_{11}), \tag{13.13}$$

where H_{11} is the 11-element of H. It is the s-zeros of expression (13) which we seek, of which there should be $4(m+p)+1$. The factor $\lambda^{2(m-1)}$ accounts for $4m - 4$ of these, all stable; it corresponds to the fact that $m - 1$ of the dynamic modes possible for the top array of oscillators cannot now be evoked. The remaining $4p + 5$ roots must be the zeros of the final bracket in expression (12). With evaluation of H_{11} this yields the equation

$$\kappa\gamma\lambda^2 + s\left[\lambda^2 + \gamma - \lambda^2 \sum_{k=1}^{p} \frac{\beta_k g_k}{\lambda^2 + g_k}\right] = 0, \tag{13.14}$$

where $G_2 = \mathrm{diag}(g_k)$ and β_k is the non-negative quantity

$$\beta_k = \gamma d^4 \sum_{j=1}^{p} r_{j1} w_{jk} r_{k1}.$$

Here r_{j1} is the $(j1)$th element of R; it is just the value of the weight r_j if the compound trace indeed takes the first of its m possible values. Expression (14) should be multiplied by $|\lambda^2 I + G_2|$ to bring it to polynomial form, $P(s) = 0$. The polynomial $P(s)$ will have a factor $(\lambda^2 + \gamma')^{b-1}$ if γ' is a non-zero b-fold eigenvalue of G_2.

Those oscillators k of the bottom array will be strongly excited for which r_{k1} is substantial. In order to get a feeling for the behaviour of the system, let us suppose that p' oscillators are strongly excited, in that for them $g_k = \gamma'$, and that the remaining $p - p'$ are unexcited, in that for them $g_k = 0$. The polynomial $P(s)$ then has a factor $\lambda^{2(p-p')}$, corresponding to the unexcited oscillators, and a factor $(\lambda^2 + \gamma')^{p'-1}$, corresponding to possible excursions of the p' active oscillators from a joint mode. This accounts for $4p - 4$ s-roots; the remaining 9 are roots of the equation

$$\kappa\gamma\lambda^2(\lambda^2 + \gamma') + s[(\lambda^2 + \gamma)(\lambda^2 + \gamma') - w\gamma\gamma'\lambda^2] = 0. \tag{13.15}$$

Here

$$w = d^4 \sum_{j=1}^{p} \sum_{k} r_{j1}w_{k1} \approx d^4|AR_1|^2,$$

where the k-summation runs over the active oscillators and R_1 is the first column of R.

Equation (15) can be tackled in various ways. One is to write it as

$$w - \gamma^{-1} - (\gamma')^{-1} = L(s)^2/(\gamma\gamma') + L(s)^{-2} + \kappa[L(s)^2 + \gamma'](\gamma's)^{-1}, \tag{13.16}$$

which we shall abbreviate to $w' = h(s)$. The locus of $h(s)$ as s describes the indented D-loop of Fig. 12.13 is sketched in Fig. 13.2, with the winding numbers marked. The direction and traverse of the various loops at infinity are determined by the fact that $h(s)$ behaves as s^{-1} near the origin and as s^4 at infinity.

We see that real w' can lie only in areas of winding number 0 or 2; i.e. there must either be stability or a limit cycle. There are areas of winding number 1 off the real axis. Whether these have any significance has yet to be determined. They correspond to the fact that, if w is complex, then what would be a pair of complex conjugate s-roots can become a pair of non-conjugate complex roots, one consistent with stability and the other not. There is no obvious physical specification which would produce such an effect. On the other hand, something like complex w could be induced if there were lags in the system (see Chapter 15) or — an effect we have not considered — if the sensory input to the system were periodic, rather than either constant or transient. These are matters to be investigated.

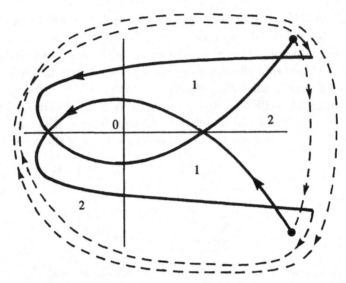

Fig. 13.2. The locus of $h(s)$, the expression in the right-hand member of (16), as s describes the indented D-loop of Fig. 12.13. The winding numbers for the various regions are indicated.

13.4. Numerical results for a two-stage system

The system in scalar variables

$$\mathcal{L}x_1 = w_1 f(x_1 - q) + f(x_2) - z_1$$
$$\mathcal{L}z_1 = f(x_1 - q)$$
$$\mathcal{D}q = \kappa[f(x_1 - q) - 0.5].$$

$$\mathcal{L}x_2 = w_2 f(x_1 - q)z_2$$
$$\mathcal{L}z_2 = f(x_2)$$

(13.17)

was solved numerically. This represents an 'escapement oscillator' system of the type of (12.28), (with variables of subscript 1), driving a second oscillator (with variables of subscript 2). This second oscillator forms a side-loop to the loop of the main oscillator rather than forming part of the main loop, so (in terms of the physiological parallels of the next chapter) could be seen as, for example, part of the PC, driven by the OB/AON loop, rather than itself part of the latter loop.

The first oscillator differs only from that set up in Section 12.9 in that it is also driven to some extent by the second oscillator; the parameters w_1 and κ are again given the values 15 and 1. Behaviour of the system was studied for a range of values of the parameter w_2, and a range of behaviour

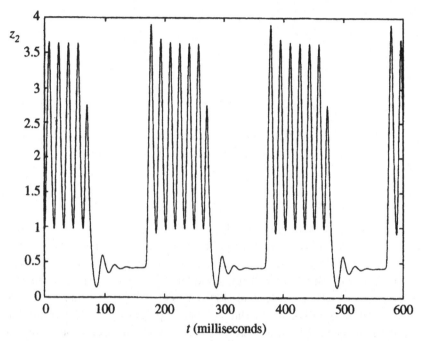

Fig. 13.3. The behaviour of z_2 for the system (17) in the case $w_1 = 15$, $= \kappa = 1$ and $w_2 = 2$. The primary oscillator (that with subscript 1) still generates the square-wave escapement oscillation of Figure 12.16; this triggers the bursts of gamma-oscillation in the secondary oscillator (that with subscript 2) so evident in the figure.

was observed. In Fig. 13.3 we give the behaviour of z_2 when $w_2 = 2$: regular bursts of 63 Hz variation, triggered when x_1 switches to the upper value of its square wave. For these parameter values the escapement oscillator shows only the simple square wave, despite the presence of the feedback term $f(x_2)$ in the first equation of (17). For other values of w_2 the first oscillator can also show some gamma-variation and the bursts of the second oscillator do not quench so cleanly.

The picture is then that block 1 (incorporating the standardising mechanism) generates a square-wave escapement oscillation as in Section 12.9, and that this at its peak triggers bursts of 63 Hz oscillation from the Freeman oscillators of block 2.

Such a structure could very well explain the bursts of gamma-waves which are such a prominent feature of EEG records. Of course, the variation is strictly periodic in this example, but would acquire a touch of chaos in a more complex system.

Chapter 14

Olfactory Comparisons.
Conclusions II

If we now wish to make some true biological connections, then the olfactory system provides a good starting point. For one thing, this is the system extensively considered by Freeman himself in his studies of both actual function and of representation by a model. For another, the system is of course an associative memory, in that it has to recognise scents, and it fits in with one of our themes, in that it must cope naturally with compound traces. The olfactory stimuli are odours, which can clearly be combined additively in a range of concentrations.

The mechanisms we have been studying are certainly broadly consistent with the operation of the olfactory system. It is well established (see e.g. Freeman (1975, 1992, 1994)) that a scent presented will produce an oscillation in the olfactory cortex, the pattern of amplitude variation over the cortex being characteristic of the scent.

The anatomy of the olfactory system is quite well understood, and is simple compared with that of the visual system, say, which must cope with data of much greater complexity and structure: with images which involve spatial relation, non-additive compounding, colour, movement and features which must remain recognisable under whole families of radical transformations. The olfactory anatomy is still complicated enough, in all conscience, but one can at least begin to marry up observed structures with those predicted from theory.

However, considered as a memory, the olfactory system does present one rather remarkable feature. It operates effectively in the *response* mode, but not in the *recall* mode. That is, an odour (or, indeed, a taste) cannot be recalled in the same vivid sense — a mental re-running of the sensory experience — as can a sight or a sound. This is a point which has long been recognised. It was, for example, made expressively by the naturalist

W.H. Hudson (1918), as he recalled the days of his boyhood on the Argentinian pampas, at that time unspoiled:

> "The psychologists tell us a sad truth when they say that taste, being the lowest or least intellectual of our five senses, is incapable of registering impressions on the mind; consequently we cannot recall or recover vanished flavours as we can recover, and mentally see and hear, long-past sights and sounds. Smells, too, when we cease smelling, vanish and return not, only we remember that blossoming orange grove where we once walked, and beds of wild thyme and penny-royal when we sat on the grass, also flowering bean and lucerne fields, filled and fed us, body and soul, with delicious perfumes."

The olfactory system certainly operates as an associative memory, in that it recognises and decomposes stimuli. It must then also operate in some degree as a storage memory, if it is to respond to a fleeting stimulus. However, it would seem that the response evoked by an actual olfactory input cannot be evoked by an act of will or even by associated inputs (e.g. the picture of an orange grove), at least not so easily or to the same degree as for the other faculties.

The capacity for dealing with composite stimuli may also be less developed than was assumed for the models of Chapter 6. It is a matter of observation that, when presented with a compound odour, one tends to be principally aware of the dominant component, even if uneasily aware that there are others. If there are two comparable strong components then attention will alternate — a subject will be aware of both components, but can give attention to only one at a time.

Perhaps compound stimuli must also be learned if they are to be recognised in their own right. This was certainly the case for the PMA, for which only those composites which had a positive prior probability could be recognised. Some remarkable correspondences are observed (necessarily for the human nose). The odour of pine-needles plus that of hydrogen sulphide ('bad eggs') comes over, in low concentrations, as grapefruit. The odour of mint plus that of rubber comes over as blackcurrant.

14.1. The anatomy of the olfactory system

If the olfactory system is simpler than that for other senses, it is still bewilderingly complex. One must remember that an animal sensory system is also coping with a whole range of other features which one would wish to blank off in a first approach, essential though they are. For example: arousal, attention, concentration, recall, sleep, consequent action, association with other sensory systems, learning and operation in a changing environment. It is then something of an achievement even to discern the grosser characteristics, let alone to explain them.

The system must at all events have an 'input' and an 'output'; we would take the first as being the receptors which provide the sensory input, and the second as being that part of the brain at which conclusions are formed, in some sense. Processing may be distributed all along the line between these, and may not be unidirectional, in that pathways almost invariably exist in both directions. (The obvious 'forward' direction is that from input to output; what we generally term 'forward' pathways are also known as 'orthodromic' or 'centripetal'. Those we term 'reverse' are also known as 'antidromic' or 'centrifugal'.) The observations which follow certainly apply to the usual experimental subjects: human, cat, rabbit and rat.

The unit into which the olfactory receptors feed after some pre-processing is the *olfactory bulb* (OB), and the part of the brain particularly concerned with olfactory information is the *prepyriform cortex* (PC). Interposed between these, however, is another unit: the *anterior olfactory nucleus* (AON). The OB in particular is represented in some detail in Fig. 14.1.

The mucosa of the nose contain some 40 million receptors (more exactly, about 10^6 in the case of humans, 10^8 in the case of dogs). These are of between 200 and 1000 types, each type responding to just one feature of an odorant molecule. The receptors are broad-spectrum, in that about 25% will react to a given odorant (although the system can also be extremely sensitive, in that stimulation of very few receptors can trigger detection and identification).

The output from these receptors is taken to the surface of the olfactory bulb by a bundle of axons, the *primary olfactory nerve* (PON). The axons from individual receptors have up to this point been kept separate, although there is some mapping, perhaps the bringing together of axons from a given receptor type. At the next stage in the bulb there is considerable reduction: the axons are led to some 2000 *glomuleri*, clusters of densely interconnected and mutually reinforcing neurons. Their internal positive feedback gives the glomuleri the character of the KI_e units described in Sections 3.4 and 12.2: an exaggerated sigmoidal response which virtually quantises their output into the two values 'high' and 'low', and also holds this state for some time.

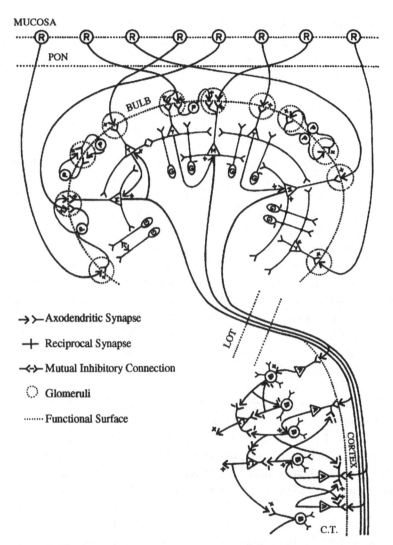

Fig. 14.1. A diagram of the olfactory system, principally of the olfactory bulb, reproduced from *Progress in Theoretical Biology*, 2, WJ. Freeman, 'Waves, pulses and the theory of neural masses', 87–165 (1972) with the kind permission of the author and of the publishers, Academic Press Inc. Excitatory and inhibitory synapses are indicated by + and −, respectively. The cell types are labelled *R* for a receptor cell, *P* for a periglomulerar cell, *M* for a mitral cell (of which a tufted cell *T* is a variant) and *G* for a granule cell. Dotted circles indicate glomuleri, which constitute assembly points for the bunching of parallel receptor inputs on to the dendrites of a set of mitral cells below. There are rich interconnections between the mitral cells.

The next stage in the bulb can be seen as a bank of Freeman oscillators, the role of excitors and inhibitors being taken by *mitral cells* and *granule cells* respectively. Complexes of these two types of cell constitute oscillators, which are also linked to each other and to other parts of the system. The mitral cells receive input from both the PON axons and the glomuleri. The mitral cells of different oscillators are also densely and symmetrically interconnected, providing a position-dependent positive feedback within the bank. There are also connections between the granule cells of different neurons, so that the inhibitors are mutually inhibiting. Accounts of the extensiveness of these connections vary; in some accounts (e.g. Yao and Freeman, 1990) they are said to be dense; in others (e.g. Baird (1986)) they are said to be short-range — so short that their effect is mediated electrically rather than axonally.

Output is taken from the mitral cells of the bulb by a bundle of axons, the *lateral olfactory tract* (LOT), to the anterior olfactory nucleus and the prepyriform cortex. To a first approximation these can also both be seen as arrays of Freeman oscillators. Freeman emphasises the internal symmetry of these arrays: a feed from one particular oscillator to another seems to be matched by a corresponding reciprocal feed. There are return pathways from the PC to the AON and the OB by the *medial olfactory tract* (MOT), and also from the AON to the OB, illustrated schematically in the block diagram of Fig. 14.5.

It seems to be generally accepted, and is emphasised by Freeman (1992), that the output from the OB to the AON and the PC is greatly condensed in the LOT, in that there are many convergences; a given PC neuron receives signal from widely-dispersed parts of the OB. There is also some divergence, in that axons originating in the OB show repeated branchings and deliver their signal to widely-dispersed parts of the PC. However, the overall effect is one of a pooling of OB outputs. Freeman sees this as achieving a function: that the spatial integration of OB activity yields a signal which expresses only the bulk mode of OB activity, so assisting synchronisation and filtering out noise. Indeed, Freeman gives the hologram analogy, that statistical information is dispersed uniformly throughout the OB and must be integrated in order to be efficiently recovered.

The feature, if real, is very surprising nevertheless. The point is often made that the dynamical state of both OB and PC is characterised more by the spatial variation of the amplitude of oscillation than by anything else. It would seem perverse, then, to lose this spatial information by a gross lumping of the OB output — a lumping that would require the most intricate temporal analysis to undo even partially.

The point is a very significant one, to which we shall return. Note, however, that the bulbocortical mapping is generally regarded as diffuse rather than degenerate, and there is at least one report of some precision in it (Scott *et al.*, 1980).

The olfactory bulb obviously achieves a considerable processing of the raw data from the receptors, but is not to be regarded as the complete associative memory on its own — for that we must look to the tri-part system of OB, AON and PC. A suggestion which we shall find ourselves led to argue in Section 4 is that the OB and the AON constitute the main stations (and termini) on the processing/communication loop between receptors and brain. The output of this loop, conveyed by the continuation of the LOT, is fed to the PC for the higher functions of interaction with other senses and activities.

Any of these three sub-systems in isolation shows a static equilibrium when unstimulated and a limit cycle when stimulated. This is also the experience with Freeman's electronic analogue of the system. Chaotic behaviour occurs in both cases first when the subsystems are coupled. Such behaviour seems to be the consequence, not merely of an increased degree of complexity, but also of the presence of delays in the system.

14.2. Neural components

It is helpful to define some particular biological sub-systems which we shall regard as standard components. Bewildering anatomical maps can then, with some licence, be reduced to block diagrams in terms of these components, and these may reveal structure and function more clearly.

One is an assembly of mutually-coupled excitatory neurons, indicated as in Fig. 14.2a. This coupling amounts to an internal positive feedback:

(a)   (b)

Fig. 14.2.   We represent the simple feedback unit of Sections 3.4 and 12.2 by the diagram (a). Feedback is achieved by diffuse interconnection of the neurons of a neural mass; the mass (a special case of what Freeman denotes by KI_e) has then essentially two excitation states. We see the glomuleri as constituting a physiological example of such a unit. An assembly of independent such units is denoted as in (b).

we suppose it rich enough that the assembly is bound into a single homogeneous dynamic system. An input is supplied in the form of a stimulus which is distributed through the assembly, and an output is taken off similarly.

This is just the structure that Freeman denotes by KI_e, a biological realisation of the simple feedback unit considered in Sections 3.4 and 12,2. It has an output which is virtually two-valued, quantising the input into 'above threshold' or 'below threshold', and also somewhat sustained. In the olfactory system we shall see it as realised by the glomuleri.

We shall denote a parallel (i.e. mutually unconnected) bank of such devices as in Fig. 14.2b. This then has a vector input and output.

Another fundamental unit is the Freeman oscillator itself: the excitatory/inhibitory pair of Section 11.5. This we shall represent as in Fig. 14.3a, the arrows at the $+$ $(-)$ end of the box representing external inputs to and outputs from the excitatory (inhibitory) component. The internal connections between the two components are understood. We see this as realised by the complexes of mitral and granule cells.

A bank of such oscillators will be represented as in Figure 14.3b. By such a 'bank' we mean a set of Freeman oscillators, mutually coupled in a manner left general for the moment, although it is understood, as ever, that the output of an excitatory (inhibitory) component is always excitatory (inhibitory) in its effect. The arrows then represent vector inputs to and outputs from the system. It is accepted that such systems are realised by the banks of mutually-connected oscillators present in OB, AON and PC. Freeman would denote such a system by KII, although he also uses the term more generally for nets simply containing neurons of both types.

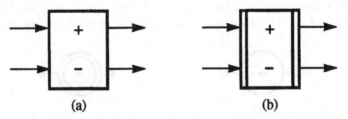

(a)   **(b)**

Fig. 14.3. We denote a Freeman oscillator by the diagram (a). The arrows at the $+$ $(-)$ end of the box represent external inputs to and outputs from the excitatory (inhibitory) component. We see this as realised by the complexes of mitral and granule cells. An assembly of such oscillators (with unspecified mutual connections) is indicated as in (b); the inputs and outputs are now vectors.

14.3. Block versions and analogues of the olfactory system

One hopes to discern some structure by reducing the complicated anatomy of Fig. 14.1 to some kind of block diagram. What must be the strongest reduction is that used by Bressler (1986), depicted in Fig. 14.4, using the conventions of Section 2. We say 'used' rather than 'proposed', because he himself probably saw it more as a provisional starting point than as even a remotely adequate representation. However, by presenting it first we can give the reader a simple picture which can then be elaborated. The OB and the PC are seen as successive banks of Freeman oscillators, the OB feeding the PC excitors by the LOT and the PC feeding the OB inhibitors by the MOT. The AON is not included.

Freeman (1975, 1987, 1992) gave the more elaborate version of Fig. 14.5, which he terms his KIII model. For all its apparent complication, it is probably the simplest which retains all significant structures and relationships. The glomuleri are represented, as is the AON (as an intermediate bank of oscillators), and the reverse paths are realised in much more detail. In particular, the AON inhibitors are excited by the PC, the OB inhibitors are excited by the AON and inhibited by the PC, and the OB excitors are excited by the AON, both directly and via the glomuleri. He also

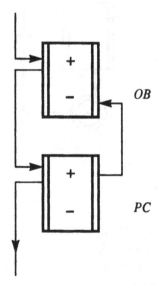

Fig. 14.4. Bressler's block diagram of the olfactory system. This is so reduced that essential mechanisms have certainly been lost, but it gives a simple picture which, once comprehended, can be elaborated to that of Fig. 14.5.

Receptors

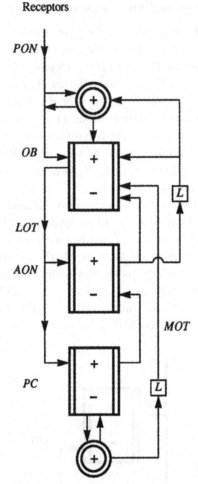

Fig. 14.5. A realistic block diagram of the olfactory system. This is essentially the version presented and simulated electronically in Freeman (1987): his KIII model and a slight simplification of the diagram given in Freeman (1975). The small blocks marked 'L' indicate a lag (i.e. a delay element).

incorporates lags (L) in the LOT and the MOT; these are physiologically realistic and certainly reinforce the tendency to chaos.

Figure 14.5 is indeed the actual plan of an electronic analogue which Freeman (1987) constructed to simulate the biological system. This was extremely successful, in that it reproduced irregular oscillations of the type observed in vivo with great faithfulness (see the two traces of Fig. 14.6).

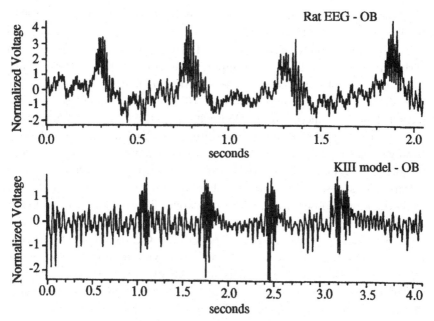

Fig. 14.6. Comparison of the EEG for the olfactory bulb of a rat with signals generated in the 'olfactory bulb' of Freeman's KIII model. Reproduced from the *International Journal of Bifurcations and Chaos*, 5, L.M. Kay, K. Shimoide and WJ. Freeman, 'Comparisons of EEG time series from rat olfactory system with model composed of nonlinear coupled oscillators', 849–858 (1995) with the kind permission of Professor Freeman and the publishers, World Scientific Publishing Co. Pte. Ltd.

It also demonstrated responses to sensory stimuli very similar to those observed in actual olfactory cortex. The analogue gave the opportunity to experiment with parameter settings. Freeman found the link at the top of the diagram, from the glomuleri to the OB, to be of particular significance. By bringing its strength up from zero one could take the system from quiescence through a sequence of bifurcations.

Each of the three sub-systems (representing the OB, AON and PC) of Freeman's analogue showed a simple periodic oscillation when isolated, the periods showing the mutual relation $T_{OB} < T_{AON} < T_{PC}$, in an obvious notation. Chaos was evident first when the three were coupled. This is broadly consistent with biological observation; Bressler (1987) notes that the OB shows periodic oscillation when isolated, the cortex shows none, but the two together (isolated from the forebrain) show more complicated variation. This perhaps indicates that the cortex is generally just sub-critical, awaiting a cue from the bulb to break into activity.

Freeman's 1987 analogue was elaborated in Yao and Freeman (1990). In this version the AON and the PC were represented as simple lumped oscillators, presumably because the LOT was regarded as providing them with an almost totally-lumped input. The only spatial pattern which could then be observed was in the separate oscillators of the analogue OB; simulations were made with 4, 8, 16 and 32 such oscillators. These indeed showed a coherent oscillation, with a spatial variation of amplitude characteristic of the stimulus (see Freeman, Yao and Burke (1988)). However, the point of some of Freeman's own physiological observations (reported e.g. in Freeman, 1994) was that such characteristic spatial variation was observed also in the olfactory cortex, and this would seem to be the crucial observation, which one would like to confirm in a model.

An incidental but interesting feature can be noted from EEG traces which Freeman presents in his 1992 paper. The OB record shows the slow breathing rhythm with a burst of 40 Hz activity at every peak. These bursts are reproduced in the AON and the PC, but the actual swell of the pulmonary cycle is not. That is, it is only the internally-generated responses, at a higher frequency, which echo around the loop.

14.4. Interpretation

The next step after forming the block diagram is to hazard some kind of interpretation, and that is, of course, the hard part. One can attempt to follow the wiring and say that 'of course' A stimulates B which stimulates C which completes *la ronde* by reinforcing A. However, such observations scarcely amount to an insight, and do not have the tight inevitability of a real explanation. One point to be kept in mind is that the oscillatory mechanisms on which we have laid such emphasis are not, after all, part of the structural logic; they are simply part of the mechanics of communication and storage. We shall look for the structural logic first; it will then be a matter of interest to see how the oscillatory mechanisms enter.

We do not venture to review the enormous BNN literature, but restrict ourselves to specific mathematical models of the ANN type.

Grossberg (1976b) regarded the OB as a standardising network of. the type represented by equations (3.17), and the PC following it as a network of type (3.18), identifying the maximal input. That is, a standardisation stage followed by what is virtually a Hamming stage. This seems too simple to be true, and the Hamming stage implies the clumsy necessity of m grandmother units, each responding to one specific trace of the m

possible. Nevertheless, if we now propose identification of the essential physiology with the PMA structure (as we shall), and believe in it (as we do), then the Grossberg proposal seems positively insightful. The PMA automatically generates the features of both a standardisation and of a Hamming net, each essential to its function. However, it also automatically generates other features: e.g. those of a feedback loop and of the two-stage structure implied by the compounding of traces. (Of course, any enlightened model, including Grossberg's later ART models discussed in Section 10.7, will employ feedback. The point of the PMA is that all these features emerge automatically and in dynamic and integrated form from the probability-maximising principle.) The standardisation, necessarily realised adaptively, also generates the slow square-wave 'escapement oscillation'.

Freeman (1992) regards the OB as a preprocessor, which quickly amplifies and codes its input so as to produce an output which can be carried on relatively few output lines, to produce the final effect of a spatial pattern of oscillation-amplitude in the brain. He is aware, if anyone is, of the role of feedback for locking the system into a dynamic whole and achieving coherent communication. Nevertheless, it is the olfactory bulb which is particularly elaborated in his models, in that it is represented by a full array of oscillators. He also makes the leap to an associative-memory function by conjecturing that the weights of the connections between the excitor elements of this array are those of a Hopfield net, in that the synaptic matrix W is proportional to AA^T (Freeman, Yao and Burke (1988), Yao and Freeman (1990), Freeman (1992)). This makes for an immediate learning rule: the Hebbian rule which, at least in the absence of observation noise, develops an estimate of AA^T.

However, one would need some justification for making this leap, particularly in view of the criticisms of the Hopfield net levelled in Chapter 6. Nevertheless, H-nets abound, and one can expect significant elements of Freeman's proposal to survive.

The formulation of a model representing reality in essential respects is a considerable ambition; before attempting it one should follow the detective-novel tradition of narrowing down the possibilities by a review of the observed facts and their implications.

First, the glomuleri, through which part of the feed from the receptors to the OB flows, appear to be KI_e structures. As deduced in Section 12.2, they would then act to increase contrast, by enhancing strong signals and depressing weak ones. They would also tend to hold their state (of 'high' or 'low' excitation) for a while, and so prolong a fleeting sensory input.

The connections between OB, AON and PC plainly serve to transmit signals between sub-systems, and to achieve this in the strong sense of locking these subsystems into a single dynamic whole. One might conjecture that it is the OB/AON loop which constitutes the principal transmission path between receptors and brain, the PC then accepting the information thus transmitted for higher processing. In support of this view, we note that the AON feeds back to the OB by three routes: directly to both the excitatory and inhibitory units of the OB, and to the excitatory units via the glomuleri. Furthermore, Freeman (1992) notes the importance of the link from this group of glomeri to the OB which emerged from experimentation with his electronic analogue. Performance of the system was extremely sensitive to the strength of the link, and lapsed entirely if this was severed.

That the AON should have leads to both the excitatory and the inhibitory elements of the OB would have the effect of feeding the signal back with a phase-shift, presumably compensating for delay in the line; and optimising phase-matching for the significant frequency band (see Chapter 15). The glomuleri would 'rectify' the signal they received from the AON, and in passing this on to the OB would reinforce and prolong the signal received from the receptors. However, note an implication of this observation: if the rectified AON signal is to reinforce then it must itself be an image (faithful up to some degree of simplification) of the signal received at this point from the receptors.

In fact, let us now boldly leap to a hypothesis, and identify the OB and the AON with the upper and lower sub-systems of the block-realisation of the probability-maximising algorithm in Chapter 8. In Fig. 14.7 we reproduce the earlier diagram of Fig. 13.1, now elaborated only to the extent of adding the holding unit H. To avoid cluttering the diagram we have not indicated the oscillator components with their separate excitatory and inhibitory feeds; if one were to do so then the lead up on the right (the analogue of the MOT) should feed into both excitatory and inhibitory ports of the upper block (the analogue of the OB).

Facile identifications are all too easy — they form the detritus through which investigators wade — but the correspondences really are quite striking. Moreover, one cannot say that they have been manufactured; the

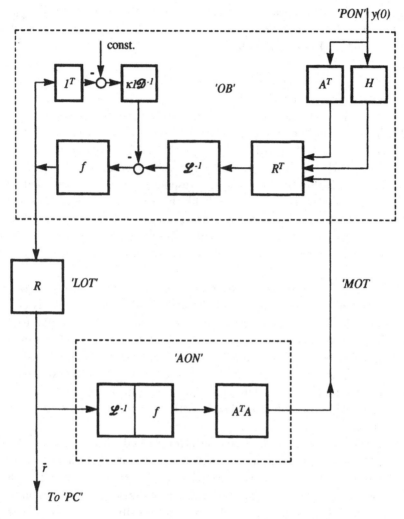

Fig. 14.7. The two-stage PMA, laid out to indicate its analogy and putative identification with the OB/AON loop. We regard the holding unit H as the analogue of the glomuleri and the upper and lower neural blocks as the analogues of the OB and the AON respectively. The loop achieves both processing and communication. The lower unit ('AON'), of dimension p, is concerned with basic trace recognition; the upper unit ('OB'), of dimension m, is concerned with recognition of the weights in which basic traces are to be combined. The forward lead incorporating the block R ('LOT') reduces the signal from dimension m to dimension p and passes it, in the purified form of an estimate of trace loadings, on to the 'PC' The oscillatory mechanisms, and so the separation of excitatory and inhibitory ports, are not indicated in this diagram; see the text.

structure of Fig. 14.7 was determined by the probability-maximising principle. Let us list these correspondences, with 'OB' for example representing the claimed analogue of the OB.

(1) The two sub-systems 'OB' and 'AON' form a natural loop, just as we have observed that the OB and AON form a strong loop, presumably achieving both transmission/synchronisation and processing.

(2) The occurrence of the product matrix AR in the PMA automatically leads to two recognisable stages in this feedback loop. One is associated with the factor A, listing the basic traces; the other with the factor R, listing the weightings corresponding to recognised compound traces. So, these respectively hold the basic trace information and form the inference on the weighting vector r. One is led to identify them with the AON and the PB respectively, the precise identification being determined by the placing of input and output lines in the PMA.

(3) Each sub-system is formed from an H-net, consistently with physiological observation. In the case of the 'AON' this is literally an H-net, in that a feed from excitor j to excitor k is balanced by an equal feed in the reverse direction. In the case of the 'OB' this is modified in that an input feed on line j to excitor k is balanced by an equal feed from excitor k to output line j. The 'OB' also incorporates the extra feature of the adaptively standardising side-loop, which one can regard as a global inhibitory effect applied directly to cell current rather than via synapses.

(4) Freeman (1992) emphasised the strongly reducing character of the LOT — an observation repeated in the literature, although at least one set of authors find evidence of some structure (Scott *et al.*, 1980). The 'LOT' of the analogue performs the operation R, a reduction from a high dimension m (the total number of possible traces) to the lowest meaningful dimension p (the number of basic traces).

(5) Furthermore, the continuation of the 'LOT' as an output would be consistent with the continuation of the LOT to the PC. The information conveyed is in its most condensed and transparent form: an estimate $\hat{r}$ of the weightings of the basic stimuli. One can arrange very easily that this evokes a corresponding spatial pattern of oscillation-amplitude in the 'PC', not needing to be recognised by a 'grandmother cell'.

(6) Inhibitory effects enter the analogue roughly as they are observed: in the 'mitral/granule' units which make up the basic oscillators, in the

global inhibition required for standardisation of output from the OB, and as an aid to phase-matching in the return loop from AON to OB.

(7) There is one feature which is peripheral but important; and on which consistency may have been achieved by construction, but by a natural one. The existence of a holding unit for the injection of data in the analogue corresponds to the layer of glomuleri on the receptor side of the OB. That part of the feed back to the OB from the AON should be directed through this is understandable: as a means of reinforcing and prolonging the essential features of the data. Part of the feed must also be direct, however, if the dynamic characteristics of the loop are not to be dampened by an intrinsically sluggish element.

The PC must of course be bound into the system by a return feed, and this will affect the dynamic character of the whole. However, one imagines this feed to be concerned more with regulation and dynamic integration than with primary transmission. In fact it is represented only by feeds to the inhibitory units of AON and OB.

It is said that no theory is useful which does not make a testable prediction. The points listed above lead us to make two; these are set out after the summary of conclusions in the next section.

14.5. Conclusions II

In this section we summarise our conclusions for Part III, as we did for Part II in Chapter 8. To represent the situation faithfully one should summarise opinions, conclusions and conjectures, but distinguish as well as one can between the three.

As far as opinion goes, we start from the admission that we are totally persuaded by W.J. Freeman's theses; in particular, the following:

(1) That relations (11.6), with the operator $\mathcal{L} = L(\mathcal{D})$ determined by (11.19), capture the essential properties of a neural mass. They then serve as a good model for such a unit, and one which is a vast improvement on the Cowan–Wilson model, valuable though this latter was as a starting point.

(2) That the pairing of excitatory and inhibitory such units, which we have termed the Freeman oscillator, constitutes a good model of the mitral/granule complexes observed to occur.

(3) That the Freeman oscillator is a fundamental BNN building block, providing a component with a threshold response and also providing

a trigger for oscillation in the wider system. It also explains central gamma-band oscillation.

(4) That the observed oscillatory activity of BNNs provides a mode of communication and state expression which is tolerant to noise, variation of baseline activity and transmission delays.

(5) That such performance is achieved despite (rather than because of) the chaotic character of these signals, which must be accepted as inevitable in a complex nonlinear system.

We shall claim as conclusions only the least arguable of assertions.

(1) That the derivation of Section 11.3 provides support for one aspect of the Freeman model of a neural mass. It explains the sigmoidal (and fairly instantaneous) nature of the response of pulse rate to dendritic current in a neural mass — by the spread in the distribution of individual neuron states.

(2) The various conclusions of Chapters 12 and 13:

  (i) That increase of positive feedback in a Freeman oscillator lowers both the stimulus threshold at which oscillation occurs and the frequency of this oscillation.

 (ii) That feedback in a system of such oscillators has much the same effect. This is of course an effect predicted by other authors with other models, and is also a matter of physiological observation.

(iii) That the incorporation of adaptive standardisation (demanded by the PMA) in a system of Freeman oscillators has a strongly stabilising effect, It also has a strongly synchronising effect, in that it produces a square wave 'escapement oscillation' which can, for example, trigger bursts of gamma waves when it peaks.

However, the conclusions on which we set most value, principally (1) below, have the character of conjectures which have the ring of truth, but not yet its certificate.

(1) That the two sub-systems of the two-stage PMA can be identified in function with the OB and the AON of the olfactory system. The grounds supporting this conjecture, listed (1)–(7) in the previous section, imply supplementary conjectures. These all concern only structural logic; the question of whether, for example, signals are or are not realised oscillatorily does not arise at this point.

(2) That the feed via the MOT to both excitatory and inhibitory units of the OB provides a phase-shift in signal which optimises loop response.

(3) That the feed from PC to OB via the glomuleri are 'rectified' in these latter and reinforce/prolong the original signal.

Finally, we make our testable predictions, necessarily with more conviction on essentials than on detail.

(a) The olfactory bulb and the anterior olfactory nucleus each contain arrays of oscillators formed from mitral/granule pairings, with synaptic connections to and from the mitral elements (excitors) of individual oscillators. It is predicted that *these connections are symmetric in the AON, in that a feed from mitral unit j to mitral unit k is balanced by an equal feed in the reverse direction.* It is predicted that, on closer examination, it will prove that *in the OB these connections are symmetric in the modified sense: that an input feed on line j to mitral unit k is balanced by an equal feed from mitral unit k to output line j.*

(b) It is predicted that *a mechanism will be found in the OB which realises the adaptive standardisation required by the PMA.* Just how this is realised is another matter, although the direct imposition of a global bias on electrical activity in the system would provide a means which is at least mathematically elegant.

Chapter 15

Transmission Delays

The question of transmission time between elements of the net is one which we have neglected entirely, despite its obvious importance in a physiological context. An investigation would be a full-dress matter which we shall not attempt here. Rather, we shall simply indicate how some of the techniques we have used extend immediately to cover the case of interstage delay. We do so largely for purposes of illustration, and shall keep to the simplest cases.

The question of delay is closely related to that of synchrony. Synchrony is certainly maintained strikingly well in BNNs, despite delays. The presence of reciprocal links certainly aids this; the possibility of effects such as the escapement oscillation may be conjectured to do so.

15.1. The Freeman oscillator with feedback

Consider first the simplest case of interest: a Freeman oscillator for which the feedback from the excitor to itself is delayed by an amount τ. A single such oscillator then amounts to a scalar version of system (12.12),

$$\mathcal{L}x = wf(\mathcal{T}x) - z + u$$
$$\mathcal{L}x = f(x),$$

with a delay effected by the lag operator $\mathcal{T}$ having the action $\mathcal{T}x(t) = x(t - \tau)$. The operator can consistently be written $\mathcal{T} = e^{-\tau \mathcal{D}}$, in that the effect of the lag is to replace w in the equivalent of the linearised perturbation equation (12.14) by $we^{-s\tau}$. Equation (12.19) then becomes

$$we^{-s\tau} = g(s). \tag{15.1}$$

Recall that $g(s) = \gamma^{-1}\lambda + \lambda^{-1}$, where γ is the gradient of f at the equilibrium value of x, and $\lambda = L(s)$. We wish to see whether equation (1)

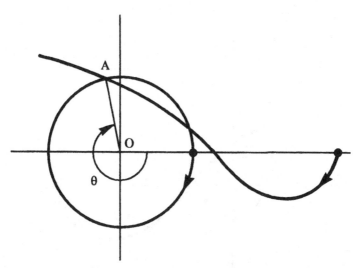

Fig. 15.1. Sketches of the loci of $g(i\omega)$ and $we^{-i\omega\tau}$ for varying ω. If an intersection point such as A corresponds to a common value of ω for the two loci then the system is on the verge of oscillation at that frequency. This will be so for the values of delay τ determined by equation (2), where θ is the angle indicated.

has solutions in $\mathrm{Re}(s) \geq 0$, and to that end examine the locus of both sides as s traces out the imaginary axis. That is, $s = i\omega$ for varying frequency ω.

We know from Section 12.4 that the locus $g(i\omega)$ is as sketched in Fig. 15.1, while the locus of $we^{-i\omega\tau}$ is a circle, centred on the origin and of radius w. The points where the two loci intersect will not in general correspond to the same value of ω; one will have to vary the parameters of the problem until they do so, if one wishes to determine threshold values of these parameters. However, one could tackle the problem in reverse, and ask for which values of τ the point of intersection *would* correspond to the same value of ω on the two loci. So, if this point is $g(i\omega_1) = we^{-i\theta}$, then this could be equated to $e^{-i\omega_1\tau}$ if

$$\tau = (\theta + 2\pi j)/\omega_1 \qquad (15.2)$$

for some integral j. Note that the angle is reckoned clockwise from the real axis. We see from Fig. 15.1 that if $w < w_\gamma$ (where w_γ is the critical value of w in the case of zero lag, determined in (12.21)) then θ must lie between π and 2π. The minimal value of the corresponding τ must then lie between π/ω_1 and $2\pi/\omega_1$. If $w > w_\gamma$ then θ can take small positive values.

Note that oscillation may occur more easily (i.e. for a smaller value of w for a given value of γ) if there is a lag of appropriate value. This is

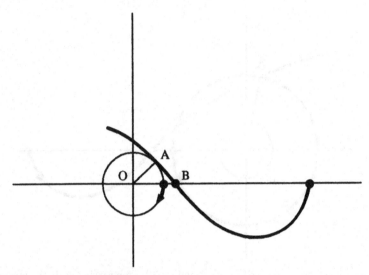

Fig. 15.2. As for Fig. 15.1. but with w chosen so that the loci meet tangentially. The fact that OA is less than OB indicates that oscillation can be evoked more easily if there is an approoriate delay than if there is none.

evident from the fact that, in the graph of Fig. 15.2, the length of the line segment OA is less than that of OB.

Freeman (1987) quotes actual lags ('latencies') between various stages of the rabbit's olfactory system as between 5 and 15 milliseconds. These are then in fact of the same order as the period (15.7 ms) of the oscillator itself. Presumably the ratio is larger in larger animals.

15.2. Interacting Freeman oscillators

A case which is somewhat closer to systems of interest is that in which two Freeman oscillators are linked to form a loop, with a transmission lag τ between them. For simplicity, we shall suppose symmetry, so the equation systems can be written

$$\mathcal{L}x_1 = wf(\mathcal{T}x_2) - z_1 + u$$
$$\mathcal{L}z_1 = f(x_1), \qquad\qquad (15.3)$$

and ditto with the subscripts 1 and 2 (referring to the two oscillators) interchanged. One readily derives the s-determining equation $(we^{-s\tau})^2 = g(s)^2$, or

$$\pm we^{-s\tau} = g(s). \qquad\qquad (15.4)$$

We are thus virtually back in case (1), with the difference that τ may now satisfy the relaxed version of (2):

$$\tau = (\theta + \pi j)/\omega_1$$

for integral j.

With the form of the feedback from the AON to the OB in mind, it would be of interest to abandon symmetry by allowing a feed from excitor 2 to *both* excitor 1 and inhibitor 1. System (3) then becomes

$$
\begin{aligned}
\mathcal{L}x_1 &= w_e f(\mathcal{T}x_2) - z_1 + u \\
\mathcal{L}z_1 &= f(x_1) + w_i f(\mathcal{T}x_2) \\
\mathcal{L}x_2 &= w f(\mathcal{T}x_1) - z_2 \\
\mathcal{L}z_2 &= f(x_2)
\end{aligned}
\tag{15.5}
$$

where w_e and w_i measure the strengths of the two corresponding feeds. The usual perturbation argument then leads to the s-determining equation

$$(\lambda^2 + \gamma_1)(\lambda^2 + \gamma_2) = w\gamma_1\gamma_2 e^{-2s\tau}(w_e\lambda - w_i), \tag{15.6}$$

where γ_1 and γ_2 are the derivatives of f at the equilibrium values of x_1 and x_2. Suppose, for simplicity, that $\gamma_1 = \gamma_2 = \gamma$. Equation (6) can then be reduced to

$$\pm e^{-s\tau}\sqrt{w(w_e - w_i\lambda^{-1})} = g(s). \tag{15.7}$$

Now, the locus of $\sqrt{\alpha - L(i\omega)^{-1}}$ has the forms illustrated in Fig. 15.3, the first or the second according as to whether α is less than or greater than $L(0)^{-1} = d^{-2}$. The effect of this factor in (7) is to distort the circular locus of $e^{-i\omega\tau}$. If we suppose that $w_e < d^{-2}w_i$ (so that we are in the first case of

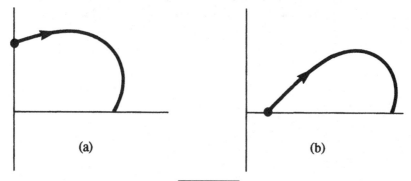

(a)   (b)

Fig. 15.3.   Sketches of the locus of $\sqrt{\alpha - L(i\omega)^{-1}}$ for varying ω. The cases (a) and (b) are those for which α is respectively less than and greater than $L(0)^{-1} = d^{-2}$.

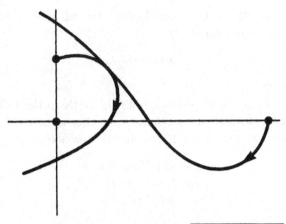

Fig. 15.4.   Sketches of the loci of $g(i\omega)$ and $e^{-i\omega\tau}\sqrt{w[w_e - w_i L(i\omega)^{-1}]}$ for varying ω. Here w_e and w_i represent feedback rates from the second oscillator to the excitatory and inhibitory ports of the first. The square-root factor rotates the locus of $e^{-i\omega\tau}$ by an amount $\pi/2$ anticlockwise and produces a bulge in it, which allows oscillation to develop at lower values of total feedback rate. Otherwise expressed, the phase shift in feedback caused by the double feed can be varied, by choice of w_e/w_i, to optimise loop response.

Fig. 15.3) and take the + option in (7), then the loci of the two sides of (7) appear as in Fig. 15.4.

What was previously the circular locus of $e^{-i\omega\tau}$ now displays a pronounced bulge as ω increases from zero. Furthermore, the locus has been rotated so that it starts from a point on the imaginary axis rather than one on the real axis. The effect is that oscillation can be realised more economically, in that the w-values can be smaller. In general, for a given τ and γ one can vary the ratio w_i/w_e in such a way as to minimise the value of $w_e + w_i$ which is required to reach threshold. That is, use of the double feed allows one to choose return phase so as to optimise loop response.

Appendix I

Extension of the Wigner Semi-Circle Law

Our aim is to prove Theorem 6.3. The classical Wigner semi-circle law holds for the case of square A (see e.g. Mehta (1991)). However, statistical applications require the analogue of this result for rectangular A. New techniques were brought to this version of the problem by Mallows and Wachter (1970, 1972) and Wachter (1974, 1978, 1980). Wachter (1974, 1978) demonstrated that, under weak conditions, the empirical eigenvalue distribution does indeed converge with increasing n to a *fixed* (i.e. non-random) distribution, and obtained (1974, 1980) the specific form of this distribution in a number of cases of statistical interest, in this Appendix we give the essentials of the Wachter/Mallows derivation of this form for the case of specific interest in this paper. It is apparent from this derivation that it is sufficient that the elements a_{ij} of A should be independent, of zero mean and unit variance and with moments of all orders.

Let the eigenvalues $U = n^{-1}A^T A$ for given n and p be λ_s ($s = 1, 2, \ldots, p$). These are random variables. Then the expectation of the Hilbert transform of the empirical distribution is

$$H_n(z) = E\left[p^{-1}\sum_s (z - \lambda_s)^{-1}\right] = \sum_{k=0}^{\infty} z^{-k-1}T_k, \qquad (A.1)$$

where

$$T_k = p^{-1}\sum_s E[\lambda_s^k] = p^{-1}E[\text{tr}(U^k)].$$

Note that the expectation E is over random variation of A; not (as in the definition of $H(z)$ before Theorem 6.3) over the asymptotic eigenvalue distribution. We can write T_k explicitly as

$$T_k = p^{-1}n^{-k}\text{tr}[(A^T A)^k]$$

$$= p^{-1}n^{-k}\sum_{ij} E[a_{i_1 j_1}a_{i_1 k_2}a_{i_2 j_2}a_{i_2 j_3}\cdots a_{i_k j_k}a_{i_k j_1}] \qquad (A.2)$$

where the summation is over the k-tuples of i and j variables.

The counting of the possible terms in this summation is eased if we represent each term by a graph in which there are nodes of two types: i-nodes and j-nodes corresponding respectively to *distinct* values of the i and j variables which occur in the a-product of (2). An a-term in the product which comes from A^T is seen as an arc from the relevant i-node to the relevant j-node; a term which comes from A is seen as an arc from j-node to i-node. The graph is then a connected graph with q i-vertices, r j-vertices and $2k$ arcs, where q and r are respectively the numbers of distinct i-values and distinct j-values which occur in the a-product of (2).

Let c_{qrk} be the number of such graphs. Then, if we carry out the summation in (2) for prescribed q and r (i.e. over permissible choices of i-values and of j-values), we obtain

$$T_k = p^{-1}n^{-k}\sum_{q,r} p^{(q)}n^{(r)}c_{qrk}$$

$$= n^{-k-1}\sum_{q,r} c_{qrk}n^{q+r}\alpha^{q-1}[1 + \mathcal{O}(n^{-1})], \qquad (A.3)$$

where $n^{(r)}$ is the combinatorial power $n(n-1)(n-2)\cdots(n-r+1)$, and we have assumed that p/n has the limit value α.

Now, the terms in the summation of (2) can be divided into three exclusive and exhaustive classes: (i) those for which at least one a-variable occurs just once, (ii) those for which all a-variables occur exactly twice and $q + r = k + 1$ and (iii) remaining terms, for which necessarily $q + r < k + 1$. The terms in the first class have zero expectation, those in the second have unit expectation and the terms in the third class give a contribution of order n^{-1} for fixed k. In the limit of large n we need then consider only the contribution from class (ii), and can reduce (3) to

$$T_k = \sum_{q+r=k+1} g_{qr}\alpha^{q-1} + \mathcal{O}(n^{-1}) \qquad (A.4)$$

where g_{qr} is the number of graphs in this restricted class. But the nodes of graphs in this class are connected either by two arcs or none at all, and, if we count the double arc as a single arc, then connectedness and the condition $q + r = k + 1$ implies that the graph is a tree. Thus g_{qr} is the number of trees which contain q i-vertices and r j-vertices and for which arcs exist only between vertices of different type. Let $\mathcal{G}$ be the class of such non-empty trees, for which then both q and r must be positive.

To evaluate g_{qr} define the generating function

$$G(\alpha, \beta) = \sum_{q,r>0} g_{qr} \alpha^q \beta^r.$$

Then this obeys the relation

$$G(\alpha, \beta) = [\alpha + G(\alpha, \beta)][\beta + G(\alpha, \beta)]. \tag{A.5}$$

To see this, break a tree on a randomly chosen arc, so that we obtain a graph of two components. Then either (i) both components are trees (of class $\mathcal{G}$, understood), (ii) one is a tree and the other an i-vertex, (iii) one is a tree and the other a j-vertex, or (iv) one is an i-vertex and the other a j-vertex. This decomposition implies relation (5) when we form generating functions.

The solution of the quadratic (5) for which $G(0, 0) = 0$ is

$$G(\alpha, \beta) = \frac{1}{2} \left[(1 - \alpha - \beta) - \sqrt{(1 - \alpha - \beta)^2 - 4\alpha\beta} \right]. \tag{A.6}$$

It follows from (1), (4) and the stochastic convergence proved by Wachter that

$$H(z) = \lim_{N \to \infty} H_N(z) = z^{-1} + \alpha^{-1} G(\alpha z^{-1}, z^{-1}), \tag{A.7}$$

and relations (6) and (7) then imply the assertion (6.16).

There are now many other proofs of this conclusion; see Opper (1989), Krogh (1992) and Sollich (1995). Sollich's proof is particularly appealing. Although not rigorous, it is short, simple and illuminating. Sollich considers the matrix $M_n = n^{-1} A^T A$ rather than $n^{-1} A A^T$ (the two have the same set of non-zero eigenvalues), and derives a recursion in p for $E[\mathrm{tr}(\lambda I - M_n)^{-1}]$, the expected Hilbert transform of the eigenvalue distribution. Conclusions follow very rapidly from this recursion and one can also see rather clearly what is required to make the argument rigorous.

Appendix II

Realisation of the PMA equation

The PMA equation

$$\dot{y} = \kappa(\theta AR\phi f R^{\mathsf{T}} A^{\mathsf{T}} y - y) \qquad \text{(A.1)}$$

appears as equation (8.18) in the text, with the exponential activation function $f(\xi) = \exp(\xi/v)$ and the standardisation factor θ defined in (8.19). This is realised (in a sense to be clarified in a moment) by the loop/block diagram of Fig. 8.3. If we premultiply equation (1) by A^{T} we obtain the analogous equation for the transformed variable $A^{\mathsf{T}} y$, which corresponds to the modified diagram Fig. 8.4.

However, there is a gap in the reasoning of section (8.5) which should be explained. Let us write equation (1) simply as

$$\dot{y} = \kappa(G(y) - y) \qquad \text{(A.2)}$$

where the sequence of operations which define the function G is represented by the sequence of blocks in the loop of Fig. 8.3. However, the loop characterises only the equilibrium version of (2),

$$G(y) = y, \qquad \text{(A.3)}$$

in which the dynamic element of equation (2) has been lost. Figure 8.3 is equivalent to the loop diagram (i) of the figure in this condensed formulation.

One could of course try to find the solution of (3) compatible with an initial value $y(0)$ of y by solving the recursion

$$y_{\tau+1} = G(y_\tau) \qquad \text{(A.4)}$$

with the initial condition $y_0 = y(0)$. This is indeed the only dynamic relation that diagram (i) could imply. Relations (2) and (4) have the same set of

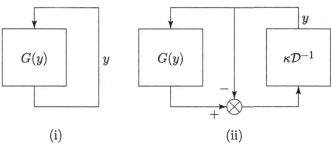

(i)   (ii)

The block diagrams (i) and (ii) represent relations (3) and (2) respectively. The symbol $\mathcal{D}$ represents the time-differential operator d/dt.

equilibria: the solutions of equation (3). However, it is not clear that each of these equilibria has the same domain of attraction in the two cases.

Diagram (ii) gives a genuine block version of the continous-time relation (2); this demonstrates the necessity for two feedback loops. When we move on to oscillatory versions of the model in Part III these points are largely taken care of by the fact that neural masses are now given explicit dynamical form. Models such as those of sections (12.4) and (12.8) then incorporate the required multiple feed-backs.

References

Amari, S. (1972) Learning patterns and pattern sequences by self-organizing nets of threshold elements. *IEEE Trans. Comput.* C-21: 1197–1206.

Amari, S. (1977a) Neural theory of association and concept formation. *Biological Cybernetics* 26: 175–185.

Amari, S. (1977b) Dynamics of pattern formation in lateral-inhibition type neural fields. *Biological Cybernetics* 27: 77–87.

Amari, S. (1980) Topographic organization of nerve fields. *Bull. Math. Biol.* 42: 339–364.

Amari, S. (1983) Field theory of self-organising neural nets. *IEEE Trans. Syst. Man. Cybernetics* SMC-13: 741–748.

Amari, S. (1988) Dynamical stability of formation of cortical maps. In *Dynamic Interactions in Neural Networks: Models and Data* (eds. M.A. Arbib and S. Amari). Springer Research Notes in Neural Computation 1: pp. 15–34.

Amari, S. (1990) Mathematical foundations of neurocomputing. *Proc. IEEE* 78: 1443–1463.

Amari, S. and Maginu, K. (1988) Statistical neurodynamics of associative memory. *Neural Networks* 1: 63–73.

Amari, S. and Takeuchi, A. (1978) Mathematical theory of formation of category detecting nerve cells. *Biological Cybernetics* 29: 127–136.

Amit, D.J. (1989) *Modelling Brain Function*. Cambridge University Press, Cambridge.

Amit, D.J., Gutfreund, H. and Sompolinsky, H. (1985) Storing infinite numbers of patterns in a spin-glass model of neural networks. *Phys. Rev. Letters* 55: 1530–1533.

Anderson, J.A. (1972) A simple neural network generating interactive memory. *Math. Biosciences* 14: 197–220.

Baird, B. (1986) Nonlinear dynamics of pattern formation and pattern recognition in the rabbit olfactory bulb. *Physica D* 22: 150–175.

Baldi, P. (1988) Linear learning: landscapes and algorithms. In *Advances in Neural Information Processing Systems* (ed. D.S. Touretzky), Morgan Kaufmann, San Mateo.

Baum, E.B., Moody, J. and Wilczek, F. (1988) Internal representations for associative memory. *Biological Cybernetics* 59: 217–228.

Beckerman, M. (1997) *Adaptive Cooperative Systems*. Wiley, New York.

Blahut, R.E. (1987) *Principles and Practice of Information Theory.* Addison-Wesley, Reading, Massachusetts.

Bressler, S.L. (1987) Functional relation of olfactory bulb and cortex. I: Spatial variation of bulbo-cortical interdependence. II: Model for driving of cortex by bulb. *Brain Research* 409: 285–301.

Bressler, S.L. and Freeman, W.J. (1980) Frequency analysis of olfactory system EEG in cat, rabbit and rat. *Electroencephalogr. Clin. Neurophysiol.* 50: 19–24.

Campbell, S. and Wang, D.L. (1996) Synchronization and desynchronization in a network of locally coupled Wilson–Cowan oscillators. *IEEE Trans. Neural Networks* 7: 541–553.

Carpenter, G.A. and Grossberg, S. (1987) A massively parallel architecture for a self-organizing neural pattern recognition machine. *Computer Vision Graphics and Image Processing* 37: 54–115.

Dayan, P. and Hinton, G.E. (1996) Varieties of Helmholtz machine. *Neural Networks* 9: 1385–1403.

Dayan, P., Hinton, G.E., Neal, R.M. and Zemel, R.S. (1995) The Helmholtz machine. *Neural Computation* 7: 889–904.

Diamantaras, K.I. and Kung, S.Y. (1997) *Principal Component Neural Networks.* Wiley, New York.

Eeckman, F.H. and Freeman, W.J. (1991) Correlations between unit firing and EEG in the rat olfactory system. *Brain Research* 528: 238–244.

Eisenberg, J., Freeman, W.J. and Burke, B. (1989) Hardware architecture of a neural network model simulating pattern recognition by the olfactory bulb. *Neural Networks* 2: 315–325.

Ermentrout, G.B. and Cowan, J.D. (1979) Temporal oscillations in neuronal nets. *J. Math. Biol.* 7: 265–280.

Freeman, W.J. (1964) A linear distributed feedback model for the prepyriform cortex. *Exp. Neurol.* 5: 525–547.

Freeman, W.J. (1972) Waves, pulses and the theory of neural masses. *Prog. Th. Biol.* 2: 87–165.

Freeman, W.J. (1975) *Mass Action in the Nervous System.* Academic Press, New York.

Freeman, W.J. (1987) Simulation of chaotic EEG patterns with a dynamic model of the olfactory system. *Biological Cybernetics* 56: 139–150.

Freeman, W.J. (1992) Tutorial on neurobiology: from single neurons to brain chaos. *Int. J. Bifurcation Chaos* 2: 451–482.

Freeman, W.J. (1994) Role of chaotic dynamics in neural plasticity. In *Progress in Brain Research,* Vol. 102 (eds. J. van Pelt, M.A. Corner, H.B.M. Uylings and F.H. Lopez da Silva), Elsevier, pp. 319–333.

Freeman, W.J. and Schneider, W. (1982) Changes in spatial patterns of rabbit olfactory EEG with conditioning to odors. *Psychophysiology* 19: 44–56.

Freeman, W.J., Yao, Y. and Burke, B. (1988) Central pattern generating and recognizing in olfactory bulb: A correlation learning rule. *Neural Networks* 1: 277–288.

Gallager, R.G. (1968) *Information Theory and Reliable Communication.* Wiley, New York.

Gardner, E. (1986) Structure of metastable states in Hopfield model. *Journal of Physics A* 19: L1047–1052.

Glendinning, P. (1994) *Stability, Instability and Chaos.* Cambridge University Press, Cambridge, England.

Goldie, C.M. and Pinch, R.G.E. (1991) *Communication Theory.* London Mathematical Society Student Texts 20, Cambridge University Press.

Grossberg, S. (1973) Contour enhancement, short term memory, and constancies in reverberating neural networks. *Stud. Appl Math.* 52: 213–257.

Grossberg, S. (1976a) Adaptive pattern classification and universal recoding. I: Parallel development and coding of neural feature detectors. *Biological Cybernetics* 23: 121–134.

Grossberg, S. (1976b) Adaptive pattern classification and universal recoding. II: Feedback, expectation, olfaction and illusions. *Biological Cybernetics* 23: 187–202.

Grossberg, S. (1982) Studies of mind and brain: Neural principles of learning, perception, development, cognition and motor control. Reidel Press, Boston.

Grossberg, S. (1988) Nonlinear neural networks: Principles, mechanisms and architectures. *Neural Networks* 1: 17–62.

Hassoun, M.H. (1995) *Fundamentals of Artificial Neural Networks.* MIT Press, Cambridge, Mass.

Hopfield, J.J. (1982) Neural networks and physical systems with emergent collective computational abilities. *Proc. Nat. Acad. Sci.* 79: 2554–2558.

Hornik, K. and Kuan, C.M. (1992) Convergence analysis of local feature extraction algorithms. *Neural Networks* 5: 229–240.

Hudson, W.H. (1918) *Far Away and Long Ago.* J.M. Dent and Sons Ltd, London and Toronto.

Kamp, Y. and Hasler, M. (1990) *Recursive Neural Networks for Associative Memory.* Wiley, Chichester and New York.

Karhunen, J. and Oja, E. (1982) New methods for stochastic approximation of truncated Karhunen-Loeve expansions. In *Proc. 6th Int. Conf. on Pattern Recognition*, pp. 550–553. Springer, New York.

Kay, L.M., Shimoide, K. and Freeman, W.J. (1995) Comparison of EEG time series from rat olfactory system with model composed of nonlinear coupled oscillators. *Int. J. Bifurcation and Chaos* 5: 849–858.

Kishimoto, K. and Amari, S. (1979) Existence and stability of local excitations in homogeneous neural fields. *J. Math. Biol.* 7: 303–318.

Kohonen, T. (1972) Correlation matrix memories. *IEEE Trans. Comput.* C-21: 353–359.

Kohonen, T. (1982) Self-organized formation of topologically correct feature maps. *Biological Cybernetics* 43: 59–69.

Kohonen, T. (1989) *Self-organisation and Associative Memory.* Springer-Verlag, Berlin.

Komlös, J. and Paturi, R. (1988) Convergence results in an associative memory model. *Neural Networks* 1: 239–250.

König, P. and Schillen, T.B. (1991) Stimulus-dependent assembly formation of oscillatory responses. *Neural Comput.* 3: 155–178.

Krogh, A. (1992) Learning with noise in a linear perceptron. *J. Phys. A.* 25: 1119–1133.

Kumar, S., Saini, S. and Prakash, P. (1996) Alien attractors and memory annihilation of structured sets in Hopfield networks. *IEEE Trans. Neural Networks* 7: 1305–1309.

Linsker, R. (1986) From basic network principles to neural architecture. *Proc. Nat. Acad. Sci.* 83: 7508–7512, 8390–8394, 8779–8783.

Lippmann, R.P. (1987) An introduction to computing with neural nets. *IEEE Magazine on Acoustics, Signal and Speech Processing* 3: 4–22.

McCulloch, W.S. and Pitts, W. (1943) A logical calculus of the ideas immanent in nervous activity. *Bull. Math. Biophys.* 5: 115–133.

Mallows, C. and Wachter, K. (1972) Valency enumeration of rooted plane trees. *J. Austral. Math. Soc.* 13: 422–476.

Mazza, C. (1997) Neural nets inference and content addressable memory. *IEEE Trans. Neural Nets* 8: 133–140.

Mehta, M.L. (1991) *Random Matrices.* Academic Press, New York. (A revised edition of the 1967 text.)

Meilijson, I., Ruppin, E. and Sipper, M. (1995) A single-iteration threshold Hamming network. *IEEE Trans. Neural Networks* 6: 261–266.

Morita, M. (1993) Associative memory with nonmonotone dynamics. *Neural Networks* 6: 115–126.

Murray, J.D. (1989) *Mathematical Biology.* Springer-Verlag, Berlin and New York.

Nakano, K. (1972) Association: A model of associative memory. *IEEE Trans. Syst. Man. Cybernetics* SMC-2: 381–388.

Nishimori, H. and Opris, I. (1993) Retrieval process of an associative memory with a general input-output function. *Neural Networks* 6: 1061–1067.

Nishimori, H., Nakamura, T. and Shiino, M. (1990) Retrieval of spatio-temporal sequence in asynchronous neural network. *Phys. Rev. A* 41: 3346–3354.

Noonburg, V.W. (1997) Threshold-dependent limit cycles in a nonlinear network model. *Neural Networks* 10: 639–648.

Oja, E. (1982) A simplified neuron model as a principal component analyzer. *J. Math. Biol.* 15: 267–273.

Oja, E. (1989) Neural networks, principal components and subspaces. *Int. J. Neural Systems* 1: 61–68.

Oja, E. (1992) Principal components, minor components and linear networks. *Neural Networks* 5: 891–902.

Opper, M. (1989) Learning in neural networks: solvable dynamics. *Europhys. Lett.* 8: 389–392.

Parisi, G. (1986) Asymmetric neural networks and process of learning. *J. Phys. A: Math. Gen.* 19: L675–L680.

Rice, J.A. (1995) *Mathematical Statistics and Data Analysis.* Wadsworth, Belmont, Calif.

Sanger, T.D. (1989) An optimality principle for unsupervised learning. In *Advances in Neural Information Prcessing Systems* (ed. D.S. Touretzky). Morgan Kaufmann, San Mateo, pp. 11–19.

Scott, J.W., McBride, R.L. and Schneider, S.R. (1980). The organization of projections from the olfactory bulb to the piriform cortex and olfactory tubercle in the rat. *J. Comp. Neurol.* 194: 519–534.

Shiino, M. (1990) Stochastic analysis of the dynamics of generalized Little–Hopfield–Hemmen type neural networks. *J. Stat. Phys.* 59: 1015–1075.

Skarda, C.A. and Freeman, W.J. (1987) How brains make chaos in order to make sense of the world. *Behavioural and Brain Sciences* 10: 161–195.

Sollich, P. (1995) Learning in large linear perceptrons and why the thermodynamic limit is relevant to the real world. In *Advances in Neural Information Processing Systems* 7 (eds. G. Tessauro, D. Touretsky and T. Leen).

Steinbuch, K. (1961) Die lernmatrix. *Kybernetik* 1, 36–45.

Strogatz, S.H. and Mirollio, R.E. (1988) Collective synchronization in lattices of non-linear oscillators with randomness. *J. Phys. A* 21: L699–L705.

Strogatz, S.H., Marcus, C.M., Westervelt, R.M. and Mirollio, R.E. (1989) Collective dynamics of coupled oscillators with random pinning. *Physica A* 36: 23–50.

Takeuchi, A. and Amari, S. (1979) Formation of topographic maps and columnar microstructures. *Biol. Cybernetics* 35: 63–72.

Taylor, W.K. (1964) Cortico-thalamic organization and memory. *Proc. Roy. Soc. B* 159, 466–478.

Thompson, J.M.T. and Stewart, H.B. (1986) *Nonlinear Dynamics and Chaos.* Wiley, Chichester and New York.

Traub, R.D., Whittington, M.A., Stanford, I.M. and Jefferys, J.G.R. (1996) A mechanism for generation of long-range synchronous fast oscillations in the cortex. *Nature* 383: 621–624.

van Hemmen, J.O. and Kühn, R. (1990) Collective phenomena in neural networks. In *Models of Neural Networks* (eds. E. Domany, J.L. van Hemmen and K. Scholten), Springer, Heidelberg.

von der Malsburg, C. (1981) The correlation theory of brain function. Internal Report 81-2, Max Planck Institute for Biophysical Chemistry. Göttingen, Germany.

von der Malsburg, C. and Buhmann, J. (1992) Sensory segmentation with coupled neural oscillators. *Biol. Cybern.* 67: 233–246.

von der Malsburg, C. and Schneider, W. (1986) A neural cocktail-party processor. *Biol. Cybern.* 54: 29–40.

von Neumann, J. (1956) Probabilistic logics and the synthesis of reliable organisms from unreliable components. In *Automats Studies* (eds. C.E. Shannon and J. McCarthy), Princeton University Press, Princeton, NJ, pp. 43–98.

Wachter, K. (1974) *Foundations of the Asymptotic Theory of Random Matrix Spectra.* Ph.D. dissertation, Cambridge University.

Wachter, K. (1978) The strong limits of random matrix spectra for sample matrices of independent elements. *Ann. Prob.* 6: 1–18.

Wachter, K. (1980) The limiting empirical measure of multiple discriminant ratios. *Ann. Stat.* 8: 937–957.

Wang, D.L. (1995) Emergent synchrony in locally couple neural oscillators. *IEEE Trans. Neural Networks* 6: 941–948.

Wang, D.L. and Ternan, D. (1995) Locally excitatory globally inhibitory oscillator networks. *IEEE Trans. Neural Networks* 6: 283–286.

Waugh, F.R., Marcus, C.M. and Westervelt, R.M. (1991) Reducing neuron gain to eliminate fixed-point attractors in an analog associative memory. *Phys. Rev. A* 43: 3131–3142.

Wehr, M. and Laurent, G. (1996) Odour encoding by temporal sequences of firing in oscillating neural assemblies. *Nature* 384: 162–166.

Whittle, P. (1962) Topographic correlation, power-law covariance functions and diffusion. *Biometrika* 49: 305–314.

Whittle, P. (1989) The antiphon, I: A device for reliable memory from unreliable elements. *Proc. Roy. Soc. London Series A* 423: 201–218.

Whittle, P. (1990) The antiphon, II: The exact evaluation of memory capacity. *Proc. Roy. Soc. London Series A* 429: 45–60.

Whittle, P. (1991a) Neural nets and implicit inference. *Ann. Appl. Prob.* 1: 173–188.

Whittle, P. (1991b) A stochastic model of an artificial neuron. *Adv. Appl Prob.* 23: 809–822.

Whittle, P. (1996) *Optimal Control; Basics and Beyond.* Wiley, Chichester.

Whittle, P. (1997) Artificial memories: capacity, basis rate and inference. *Neural Networks* 10: 1619–1626.

Wilson, H.R. and Cowan, J.D. (1972) Excitatory and inhibitory interactions in localized populations of model neurons. *Biophys. J.* 12: 1–24.

Winfree, A.T. (1967) Biological rhythms and the behaviour of populations of coupled oscillators. *J. Theor. Biol.* 16: 15–42.

Xu, Z., Hu, G.Q. and Kwong, C.P. (1996) Asymmetric Hopfield-type neural model: Theory and applications. *Neural Networks* 9: 483–501.

Yao, Y. and Freeman, W.J. (1990) Model of biological pattern recognition with spatially chaotic dynamics. *Neural Networks* 3: 153–170.

Yoshizawa, S., Morita, M. and Amari, S. (1993) Capacity of associative memory using a non monotone neuron model. *Neural Networks* 6: 167–176.

Index